Dosimetry

Edited by Thomas J. FitzGerald and Maryann Bishop-Jodoin

Published in London, United Kingdom

Dosimetry
http://dx.doi.org/10.5772/intechopen.98044
Edited by Thomas J. FitzGerald and Maryann Bishop-Jodoin

Contributors
Wan Shun Leung, Hing Ming Hung, E. Ishmael Parsai, Elahheh Salari, Carlos Eduardo De Almeida, Camila Salata, Hossam Donya, Maria E Lyra-Georgosopoulou, Dimitris A. Verganelakis, Thomas J. FitzGerald, Linda Ding, Carla Bradford, Kenneth Ulin, Koren Smith, I-Lin Kuo, Yankhua Fan, Abdulnasser Khalifeh, Fenghong Liu, Harry Bushe, Camelia Bunaciu, Jonathan Saleeby, Shannon Higgins, Julie Trifone, Maureen Britton, Joshua Taylor, Marious Croos, Katie Figura, Thomas Quinn, Salvatore LaRosa, Linda Martin, Kathleen Briggs, Sherri Suhl, Jean Quigley, Heather Reifler, Shawn Kirby, Fred Prior, Joel Saltz, Maryann Bishop-Jodoin, Suhong Lu, Said Elboukhari, Khalid Yamni, Hmad Ouabi, Taoufiq Bouassa, Lahcen Ait Mlouk, Payal Raina, Rashmi Singh, Mithu Barthakur

First published in London, United Kingdom, 2022 by IntechOpen
IntechOpen is the global imprint of INTECHOPEN LIMITED, registered in England and Wales, registration number: 11086078, 5 Princes Gate Court, London, SW7 2QJ, United Kingdom

British Library Cataloguing-in-Publication Data
A catalogue record for this book is available from the British Library

Additional hard and PDF copies can be obtained from orders@intechopen.com

Dosimetry
Edited by Thomas J. FitzGerald and Maryann Bishop-Jodoin
p. cm.
Print ISBN 978-1-80355-459-4
Online ISBN 978-1-80355-460-0
eBook (PDF) ISBN 978-1-80355-461-7

Meet the editors

Dr. Thomas J. FitzGerald is a professor and chair of the Department of Radiation Oncology, UMass Chan Medical School, USA. The department has a strong academic mission and provides clinical care to multiple communities throughout central Massachusetts with several community centers providing advanced technology clinical care. Dr. FitzGerald has been the principal investigator of the Quality Assurance Review Center (QARC) for more than twenty-five years. QARC provides imaging and radiation oncology data acquisition and data management services to the National Clinical Trials Network (NCTN) and industry partners. QARC is now part of the Imaging and Radiation Oncology Core (IROC), which centralizes data management and quality assurance service for NCTN clinical trials with offices in Rhode Island, Houston, TX, Columbus, OH, and Philadelphia, PA. In this capacity, proton institutions' applications are credentialed for clinical trial participation and data reviewed for protocol compliance. Dr. FitzGerald serves in an advisory capacity for The Cancer Imaging Archive (TCIA).

Maryann Bishop-Jodoin, MEd, is an instructor and scientific writer in the Department of Radiation Oncology, UMass Chan Medical School, USA. She has worked with her co-editor Dr. FitzGerald at the Quality Assurance Review Center for more than twenty-five years. She writes and edits manuscripts, books, grants, and other publications.

Contents

Preface

Modern dosimetry requires many different tools including skills in mathematics and science. The skill set for the modern treatment planning team has matured to include multiple areas of expertise, including planning in four dimensions, motion management, intensity modulation, adaptive planning, brachytherapy, and theranostics. Schools and teaching programs are constantly adapting to meet the needs of modern physics and dosimetry teams. Modern clinical practice programs need to support students at this level to better prepare students for the changing work environment. In this book, we want to touch on these points and identify what is needed for modern patient care.

The book describes multiple concepts and challenges of modern dosimetry. As treatment planning systems have become more commercialized and dose computation systems have become more uniform, a different scope of challenges faces modern physics and dosimetry teams participating in patient care at all institutions. The chapters in this book address an array of clinical challenges including processes in pre-clinical plan and dose validation, reference dosimetry, thermoluminescence, modern issues affecting pediatric populations, comparative dosimetry, including ^{60}Co and ^{192}Ir, modern applications of dosimetry relative to intensity modulation, and current challenges in therapy planning including image applications for planning radiation therapy in all disease areas.

Our goal is to shed light on problem-solving strategies from a diverse group of practitioners and move toward a more generalized understanding of how issues are addressed by global colleagues. Modern patient care protocols will require worldwide participation as we work to understand the roles of local and systemic therapy on diverse populations. More comprehensive standardization in our pretreatment and treatment processes will decrease and potentially eliminate variables in data interpretation. In this book, we describe the current status of dosimetry standards as we work towards uniform application of dosimetry in developing consistent treatment planning standards. The role of imaging in this effort cannot be understated, as modern imaging will both define standards for contour as well as serve to validate radiation dose.

We appreciate the opportunity to present this effort to you.

Thomas J. FitzGerald MD and Maryann Bishop-Jodoin
Department of Radiation Oncology,
University of Massachusetts Medical School,
Worcester, MA, USA

Chapter 1

Clinical Considerations for Modern Dosimetry and Future Directions for Treatment Planning

Linda Ding, Carla D. Bradford, Kenneth Ulin, Koren Smith, I-Lin Kuo, Yankhua Fan, Abdulnasser Khalifeh, Fenghong Liu, Suhong Lu, Harry Bushe, Salvatore Larosa, Camelia Bunaciu, Jonathan Saleeby, Shannon Higgins, Julie Trifone, Maureen Britton, Joshua Taylor, Marious Croos, Katie Figura, Thomas Quinn, Linda O'Connor, Kathleen Briggs, Sherri Suhl, Jean Quigley, Heather Reifler, Shawn Kirby, Fred Prior, Joel Saltz, Maryann Bishop-Jodoin and Thomas J. FitzGerald

Abstract

Technology and computational analytics are moving forward at an extraordinary rate with changes in patient care and department workflows. This rapid pace of change often requires initiating and maintaining the educational support at multiple levels to introduce technology to radiation oncology staff members. Modern physics quality assurance and dosimetry treatment planning now require expertise beyond traditional skill based in computational algorithms and image management including quality assurance of the process of image acquisition and fusion of image datasets. Expertise in volumetric anatomy and normal tissue contouring are skills now performed by physics/dosimetry in collaboration with physicians and these skills are required in modern physics dosimetry training programs. In this chapter, challenges of modern radiation planning are reviewed for each disease site. Skills including future applications of image integration into planning objects and the future utility of artificial intelligence in modern radiation therapy treatment planning are reviewed as these issues will need to be added to modern training programs.

Keywords: radiation treatment, artificial intelligence, clinical trials, quality assurance, treatment planning, image-based volumetric dosimetry, outcome

1. Introduction

Prior to the advent of volumetric radiation oncology treatment planning, physician, physics, and dosimetry teams would construct and calculate radiation therapy treatment plans at the center of the target referred to as the isocenter. Calculations would be derived based on depth measured at isocenter. Beam shaping devices which shaped dose at the isocenter were applied to the sloping surface of the target at a single level. Plans would be calculated to isodose lines which would unintentionally not define the volume and location of areas of radiation dose asymmetry. In breast cancer patients, the areas of asymmetry would be at the medial and lateral regions of the breast in rib/chest wall and extend for the length of the field which by default would include multiple rib segments. In this era injury, when it occurred, was simply ascribed to radiation therapy with limited attention to dose and volume treated. Radiation is not a drug; however, we did not have volumetric computational tools to be more exact in our review of process and convince our medical and surgical colleagues otherwise. We could not determine a specific dose volume effect relative to injury as we did not have tools to validate this point. Our approach to treatment planning changed with the introduction of computer tomography into simulation and volumetric driven radiation therapy treatment planning. With tools for contouring targets with reconstruction algorithms, radiation oncologists and treatment planning teams could visualize targets in three and four-dimensional volumes and review the juxtaposition of target volumes with normal tissue structures. This provided radiation oncologists opportunities to apply therapy in non-coplanar modulated geometries with beam arrangements that were more specific to each patient's target volume.

Radiation therapy treatment planning and dose prescription permanently changed with the introduction of advanced technology. Dose was prescribed relative to volume, not isodose lines, and contouring normal tissues provided the infrastructure to develop strategies for conformal avoidance of normal tissue. Altering fluence profiles by moving multi-leaf collimators across radiation therapy treatment fields provide the opportunity to generate voxel-based dosimetry to further improve dose asymmetry to tumor targets and place sharper dose gradients across normal tissue targets decreasing the risk potential for injury. The weakness in target specificity was therapy reproducibility and image validation of the targets before and during therapy. This was addressed with several manufacturing improvements including the integration of cone beam computer tomography into linear accelerators, use of four-dimensional treatment planning to develop therapy targets, and optical tracking tools to validate the lack of motion during treatment. Volume modulated arc treatment delivery provided the opportunity to decrease the time of treatment delivery with simultaneous multi-leaf and gantry motion. By decreasing treatment time delivery both durability and reproducibility of daily positioning could be confirmed. Motion, including deep inspiration breathing, is now validated with optical tracking systems. Decreasing the time of treatment delivery with volume modulated arc therapy provides confidence that patient care will not be influenced by intra-fractional motion.

These improvements have served to secure the success of radiation oncology moving forward as tumor targets are treated accurately with confidence and normal tissue protection is optimized. The improvements have also served to change the work scope and skill set of colleagues in the radiation therapy physics and treatment planning community. With the advent of volume-based planning, image integration into targeting has become the standard of care. With two-dimensional radiation therapy treatment planning, often the information needed to generate a standard treatment plan was fully available to the dosimetry and physics teams at the end of the simulation hour. Today, most of the work in planning and targeting is performed after the simulation process. The simulation hour is used to create devices for immobilization, perform three- and four-dimensional imaging, and establish target coordinates for planning. Physicians contour targets for treatment after images are acquired and processed, often with diagnostic images fused into radiation therapy planning images. The work of the planning team cannot begin in earnest until the targets for therapy are contoured and constraints for normal tissues are defined for the objectives of the planned therapy. If there are delays in physician contouring, unintentional time constraint can be placed onto the planning and treatment validation process. The planning teams need to be well versed in volumetric therapy language as clinical, motion, and planning targets are applied to the intended areas of therapy by planning teams following protocol and/or institutional guidelines. The plan is developed as best as possible within the confines of the normal tissue volume constraints and validated through the quality assurance process. Image guidance and tracking process for quality assurance is initiated and maintained by the planning teams in collaboration with the therapy teams. Planning teams are essential in all services housed within the department of radiation oncology from the time of simulation to treatment validation. Planning teams are involved in brachytherapy and stereotactic therapy with varied imaging and dose computational algorithms required for modern patient care. The skill set for the modern planning team is diverse requiring knowledge of all aspects of modern planning equipment and tools.

Therefore, the role of the planning team in radiation oncology has expanded to image-based volumetric dosimetry and plan validation. Dosimetrists now have an extended role in defining volumetric anatomy and plan validation. In this chapter we will review skills required by dosimetry and planning teams in each disease and discipline area; the role of imaging and dosimetry in both daily work scope and clinical trials; the skill set for the modern planning team; and define what a modern planning group might resemble soon [1–11].

2. Central nervous system

Patients requiring radiation therapy to the central nervous system (CNS) require a broad range of department services as these patients comprise both pediatric and adult populations including patients with primary and metastatic disease. Patients can require therapy to the entire CNS as well as stereotactic therapy to small targets with sharp dose gradients across critical normal tissues. The objectives for each patient have similarities with the primary goal to successfully treat the tumor target with conformal avoidance to as much central nervous parenchyma and critical structures as feasible and not compromise dose to target. In both pediatric and adult populations, sparing normal tissue now has near equivalent importance in patient management to tumor control and this has become essential to the treatment planning community. The CNS has multiple critical normal tissue structures with limited self-renewal capacity, therefore conformal avoidance when possible is important for optimal clinical outcome. Imaging plays an essential role in defining targets and accurate contouring is fully dependent on image fusion and quality. Most tumor structures are not well visualized on computer tomography. Fusion software is aligned with bony anatomy and the irregular shape of skull landmarks lends itself to accurate integration of multiple image sets for contouring. Investigators have developed protocols in glioblastoma multiforme using multiple magnetic resonance (MR) sequences involving spectroscopy, fluid-attenuated inversion recovery (FLAIR), and contrast images using dose painting strategies to help limit dose to critical structures. Spectroscopy is helpful when tumor abuts the corpus callosum and can define areas where disease extends to the contralateral hemisphere, FLAIR defines edema which can house disease, and contrast defines regions of blood-brain barrier disruption by disease. Positron tomography imaging with amino acids can define tumor in deoxyribonucleic acid (DNA) synthesis often not well visualized with gadolinium. The datasets help create multiple target volumes which can be treated as a single plan with individual fractionation and total dose to each target. For patients with metastatic disease, treatments are delivered with radiation treatment planning including radiosurgery to subtotal volume CNS targets and hippocampal sparing for improved neurocognition. The growth of MR imaging has facilitated the development of subtotal volume therapy. Pediatric radiation therapy on selected germ cell protocols is delivered to spinal fluid pathways with temporal lobe sparing and standard risk medulloblastoma therapy boosts are now planned to image targets and not the entire posterior fossa in order to spare normal tissue. The plans require creativity with a balance of constraints between multiple normal tissue targets with dose limitations applied to the CNS tissue volume in the treatment field.

Often tumor targets come in close approximation to normal tissue and planning teams need to be fluent in multiple aspects of field geometry including table motion, off-axis fields, and six-degree couch motion and place dose gradients across structures including optic nerves, chiasm, and the cochlea when needed. Artificial intelligence (AI) will have influence in this aspect of care as field design can be optimized to constraints through an iterative process once the contours have been drawn and processed. The plan, once approved, is validated through a quality assurance process

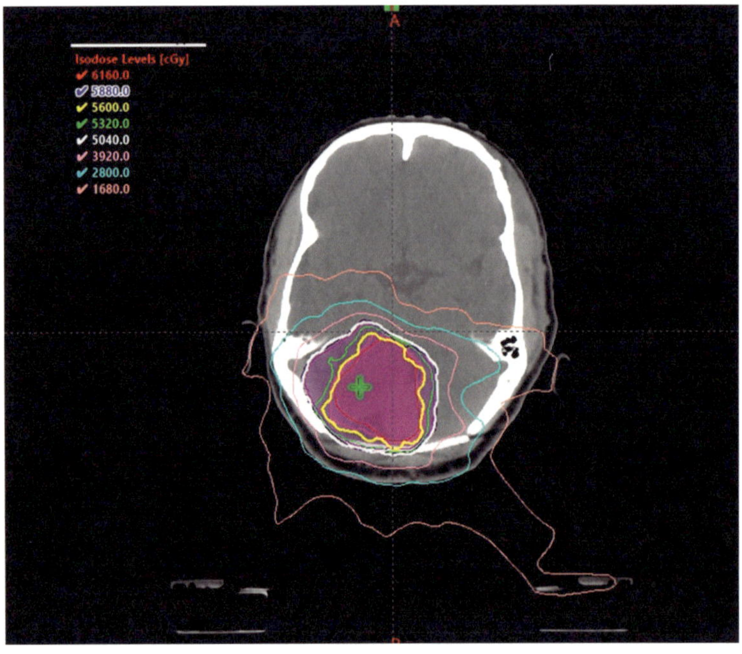

Figure 1.
Dose painting intensity-modulated radiation therapy (IMRT) plan for a posterior fossa glioma.

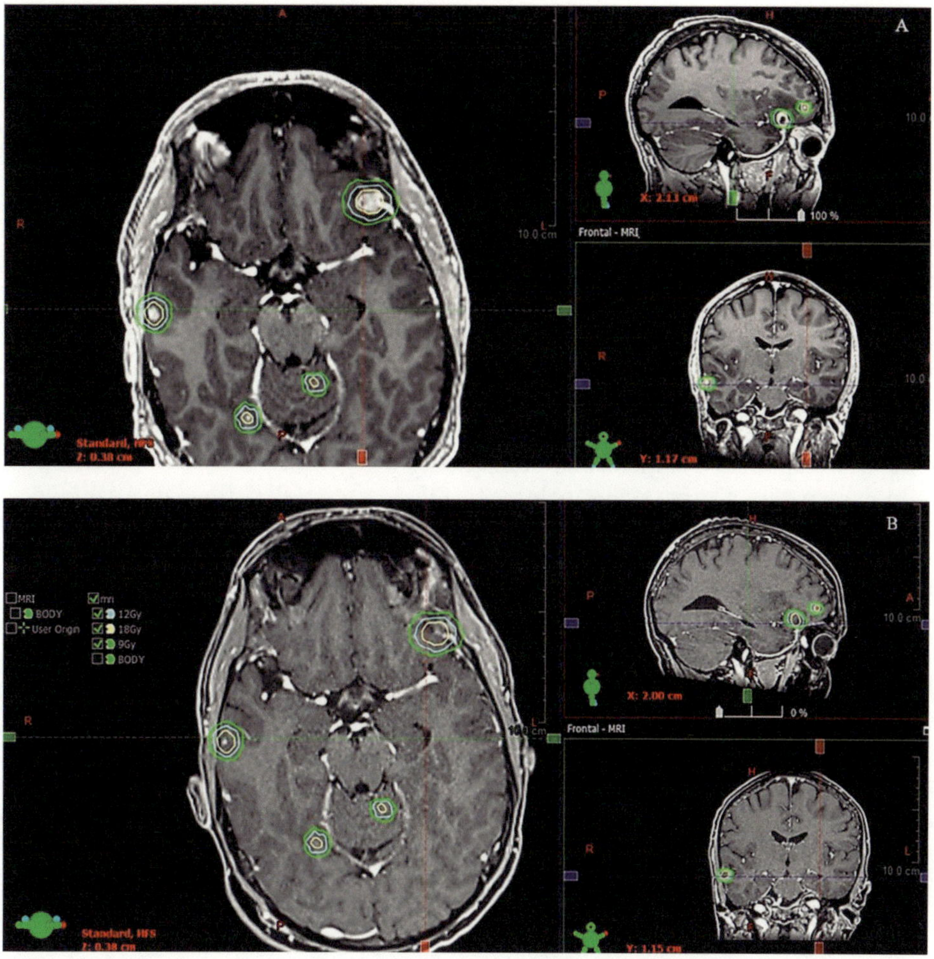

Figure 2.
Dose delivery to multiple lesions with a mono-isocenter in the central nervous system using a single plan to treat all lesions. (A) Pre-SRS MRI; and (B) Isodose lines (Rx = 18Gy to 11 tumors in single fraction treatment) on post treatment MRI 8 months later.

and treatment can begin once the patient's therapy portal images are generated and approved. **Figure 1** is the treatment plan of a patient with neurofibromatosis with an astrocytoma in the posterior fossa occupying the fourth ventricle showing high and intermediate risk volumes defined on MR with conformal avoidance of the cochlea.

Planning for diseases in the CNS is clinically important as normal tissues of the CNS have limited self-renewal potential, therefore conformal avoidance to as many structures as possible with radiosurgery (SRS) and stereotactic radiation therapy is essential for outcome. Few diseases alter the well-being of the patient more than injury to the CNS imposed by disease and treatment. Limiting sequelae of therapy is an essential goal for the dosimetry team [12–23]. Further improvements in small field dosimetry permit multiple lesions to be treated in a segmental manner with a single plan with one isocenter. **Figure 2** is an example of a single isocenter plan simultaneously treating multiple lesions in the CNS using the Varian RapidPlan system with volume modulated arcs. The arc permits simultaneous dynamic motion of the treatment gantry and multi-leaf motion to optimize delivery to tumor and limit dose to normal tissue.

3. Head/neck

With the advent and recognition of human papilloma virus (HPV), the incidence of head and neck carcinoma has significantly increased including patients who are lifelong non-smokers. These are challenging patients to plan as there are a multitude of normal tissues to provide both conformal avoidance and titration of dose asymmetry. Our knowledge of dose constraints continues to mature and we are applying strategy to as many normal tissue volumes as the primary target will permit. Spinal cord dose is limited to 50 Gy to 1% when feasible. Dose volume constraints are applied to the mandible, retropharyngeal muscles, carotid vessels, larynx, brachial plexus, and thyroid when possible. When tumors involve the paranasal sinus and skull base, constraints need to be applied to orbital structures, pituitary gland, optic chiasm, cranial nerves and temporal lobe including the hippocampus when possible. These constraints are balanced with tumor location and normal tissue anatomy coupled with knowledge of tissue molecular biomarkers.

In these patients, gross tumor volume (GTV) is often well defined on anatomic and metabolic imaging coupled with findings on physical examination. This permits more expert targeting of disease location as well as clinical target volumes (CTV) thought to be at risk for disease, often considered one lymph node station beyond gross tumor. Planning target volumes (PTV) are designed for daily set up variability, however with the advent of optical tracking tools and improved image guidance, the traditional need for a 5 mm PTV can be re-visited as arbitrary application of expanded targets can often extend target dose into normal tissue structures and outside of the patient if not carefully applied. Vertebral bodies can often be natural barriers especially if volume expansion places the spinal cord at risk for additional radiation dose. Modern image guidance has likewise improved daily reproducibility of radiation therapy treatment. This allows departments to re-visit the concept of a PTV

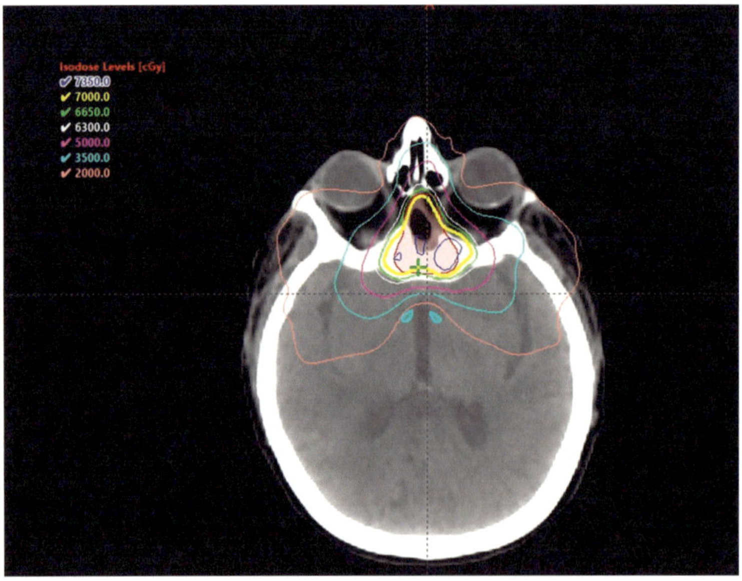

Figure 3.
IMRT plan for ethmoid sinus esthesioneuroblastoma. Note conformal avoidance of the globes.

since daily patient treatment is more secure and consistent. Titrating the volume will decrease radiation dose asymmetry and improve clinical outcome.

In applying constraints, it will be important for therapy teams to track outcome through pathways previously less well studied. For example, although it is likely that the dose/volume relationships of retropharyngeal tissue treated influences outcome, recognizing this is driven in large part by the location of primary disease. We need strategies including speech/swallowing colleagues to study this effect and learn where to dose/volume reduce when feasible. Audiologists will help radiation oncologists apply metrics to outcome for alterations in hearing imposed by therapy. Building a portfolio for outcome analysis will support process improvements in radiation therapy planning and support the identification of patients who would potentially benefit from supportive intervention prior to the appearance of visible sequelae of management. **Figure 3** represents the plan of a patient with recurrent paranasal sinus esthesioneuroblastoma with conformal avoidance of the optic chiasm.

This is a growing population of patients who will benefit from the attention to detail required for optimal tumor control and normal tissue function [21, 24–30].

4. Thorax

Lung cancer has evolved over the past two decades from a disease of habit to a disease of molecular biology. Lesions are now treated with multiple techniques including discontinuous planning such as dose painting/altered fractionation to peripheral nodules and more traditional fractionation to regions of central mediastinal disease where tumor abuts critical central structures. With the advent of immunotherapy, the situation has become more complex as toxicity can occur in both high and low dose regions making planning and dose constraints challenging to meet. Lung tumors can be large and often are in less favorable thoracic locations to meet cardiac and pulmonary constraints. Accordingly, modern lung cancer protocols that include immune-radiotherapy place strict constraints on V20 and V5 with a dose ceiling of 60 Gy. To meet constraints, nearly all modern industry and National Cancer Institute's (NCI) National Clinical Trials Network (NCTN) clinical trials treat limited to no elective at risk volumes. Only gross tumor as defined on anatomic and metabolic imaging is often contoured with dose constraints placed on cardiopulmonary volumes, soft tissue, chest wall, and spinal cord.

Thoracic planning will remain an enigma with imperfections applying dose to structures. Most thoracic plans are now performed with intensity modulation. If left to its own device, intensity modulation will titrate dose asymmetry (hot spots) but unfortunately push dose to vulnerable pulmonary parenchyma if a strict low dose (V5) and moderate dose (V20) constraint is applied. When these constraints are applied, dose will be driven back to high dose segments which in turn will create hot spots, largely in anterior/posterior soft tissue intentionally lateral to the spinal cord. This has the potential of increasing dose to the chest wall and soft tissue structures. Often this is viewed as acceptable in order to prioritize more limited dose to pulmonary parenchyma. This results in the need for balance of constraints.

As the number of cancer survivors grow [31], the modern cancer patient is asking different questions concerning outcome beyond the question of tumor control. Thoracic malignancies including primary lung, esophagus, and lymphoma can generate dose to vascular structures and therefore cardiac dosimetry is an important element to thoracic therapy. Historical therapy directed to the heart as an unintentional target is associated with coronary artery disease, myocardial dyskinesis, valvular disease, electrical conduction changes, and pericardial disease. It is now also recognized that as blood migrates through chambers during treatment, radiation can decrease white cells and other blood elements. Once thought to be exclusively related to marrow dose, the heart becomes a vehicle for immune suppression through therapy. Therefore, for all epithelial and liquid diseases of the thorax, attention to detail to the heart and cardiac subsegments need to be assigned constraints when feasible. This information becomes invaluable to both the cardiologist and primary care provider teams in evaluating the patient and creating a survivorship plan including cardiac prevention strategies. We cannot evaluate radiation as a "drug" and to optimize survivorship programs, defining dose to subsegments and vessels will provide meaningful information to patients and care providers. Future AI tools will help optimize consistent application of contours to subsegments which will optimize strategies for conformal avoidance by the planning team. The esophagus abuts the left atrium; therefore, radiation oncologists can provide information on dose to the electrical conduction system as a cardiac subsegment as part of survivorship planning as tumor target will abut the posterior wall of the atrium. Improvements such as this instill confidence in providers and patients recognizing that we place value on outcome. The same approach can be adopted to other normal tissue volumes including pulmonary parenchyma. Often thoracic lung cancer patients have compromised baseline function with limited pulmonary capacity, therefore conformal avoidance of parenchyma is important.

Future strategies for application of tools for planning will include functional coefficients for cardiopulmonary volumes. Currently we can only assign anatomical coefficients without recognizing that function may be an important component to future planning paradigms. Lung will remain a balance of constraints including low/intermediate, and high dose parenchymal segments as well as cardiac, esophageal, spinal cord, the chest wall volumes. Integrated databases with images, radiation therapy plans, and outcome with help us further refine planning strategies for thoracic oncology patients (**Figure 4**) [32–63].

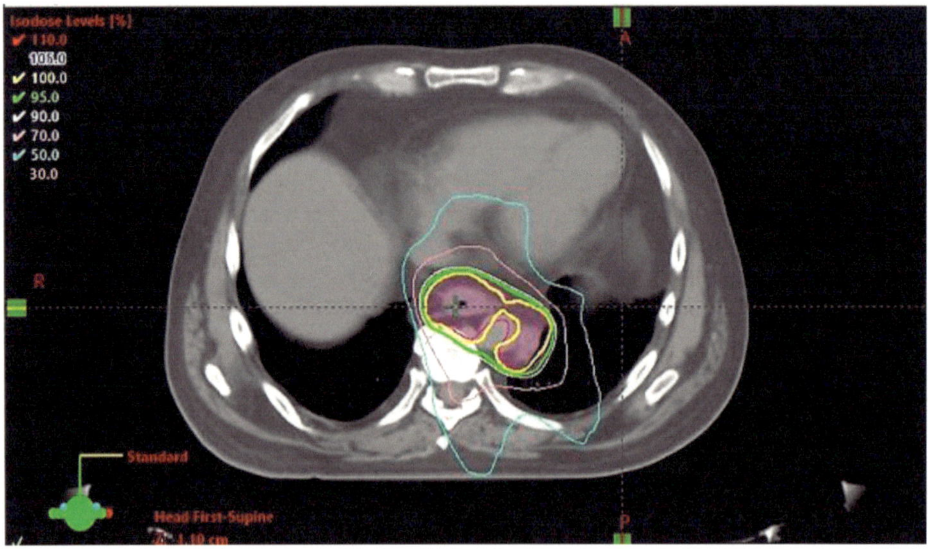

Figure 4.
Esophagus patient with a pre-operative radiation therapy treatment plan demonstrating conformal avoidance of cardiac and pulmonary volumes.

5. Abdomen/pelvis

Radiation therapy plays an important role in the treatment of abdominal and pelvic malignancies which include the need for stereotactic therapy and brachytherapy. This requires an expanded skill set for the planning team as the team must prepare for additional therapy superimposed on teletherapy.

In the past two decades the liver has become an important focus for radiation therapy. Systemic therapies have improved for metastatic disease and primary hepatocellular carcinoma has significantly increased in incidence; therefore, radiation oncologists are applying advanced technology tools to the management of these patients including stereotactic therapy for definitive management and as a bridge to transplant. Planning teams become fluent in fusion of multiple MR and metabolic image sets into planning computer tomography including motion management techniques for successful treatment planning and delivery. Liver targets are often difficult to visualize in early iterative versions of cone beam computer tomography to validate target positioning for therapy, therefore planning teams have used versions of fiducial tracking to validate positioning for treatment set up. Multiple constraints are applied to liver targets including mean dose and partial volume dose. Constraints are applied to gastric/small tissues in close approximation to the liver as well as cardiac, pulmonary, and renal constraints applied in selected areas relative to target volume location. This is an important area of clinical research as radiation therapy is also being supplied via radio pharmacy with Y-90 and other compounds in development. Additional theranostic therapies have hepatic uptake, therefore this dose will need to be calculated as well for dose volume analysis.

Volumetric dosimetry is needed in this area. Although Y-90 can be applied to the region of disease, tumor vascularity may prevent uniform application of dose and the area of intended therapy may be underdosed with migration of therapy away from the intended target. The strategy for volumetric dosimetry with both diffusion kinetics and evaluation of migration is being developed and will likely include serial single positron computer tomography images obtained frequently (daily) and fused into a planning computer tomography to evaluate areas of disease potentially undertreated which can then be augmented with stereotactic therapy. Likewise, areas of dose migration can be identified as regions of conformal avoidance for the radiation planning team through these processes. Modern planning teams need to be nimble in image fusion and registration in order to optimize patient care in this group of patients. **Figure 5** represents a stereotactic hepatic radiosurgery treatment.

Upper abdominal therapy is often directed to biliary, pancreatic, gastric, and lower esophageal targets. In this cohort of patients often bowel, renal, liver, and occasionally cardio/pulmonary constraints need to be applied with thought. In patients treated in a post-operative setting, tolerance to bowel may need to be titrated, especially if it is devascularized and fixed in position. This is commonly seen after pelvic surgery. Likewise, the regions of the gastrointestinal and biliary anastomoses must be identified for conformal avoidance/dose titration as best as possible. Invariably, these areas can reside in high-risk regions, therefore advanced planning techniques need to be applied to optimize patient care and avoid injury.

For treatment to the pelvis, often bowel, bladder, and rectum are considered targets at risk. Mucosal surfaces of these organs are tissues of self-renewal potential and injury is often related to limitations in cell re-growth along the mucosal surface resulting in bleeding/nerve exposure resulting in pain. Long term effects such as

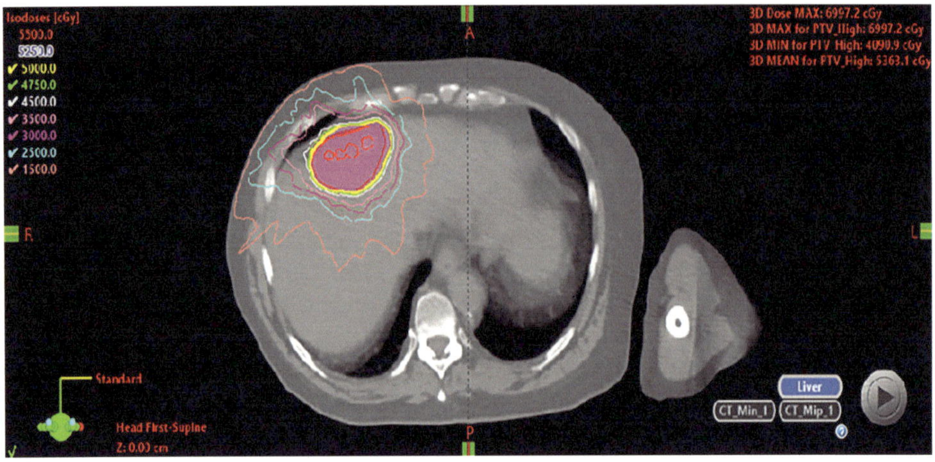

Figure 5.
Hepatic stereotactic body radiosurgery plan.

perforation are unusual appear to center of areas of previous surgical intervention which are inherently devascularized/fixed in position. This is true for all surgical colleagues including gynecological oncologists, urologists, colorectal surgeons, and surgical oncologists. When possible, these areas need to be identified pre- radiation therapy for dose titration when not intentionally included in high-risk tumor volumes. Although we can treat more tissue than a surgeon can remove including extended nodal volumes, outcomes in post-operative patients require careful exchange between the radiation oncologist and surgical colleagues in designing target volumes of interest to optimize patient care and place dose gradients across tissues considered vulnerable to injury.

There is increasing interest in the use of radiation therapy for abdominal and pelvic malignancies and use of stereotactic techniques when feasible to limit sequelae of management. The modern physics team will increasingly use advanced technology in the care of these patients. As we improve our technology, we must be cognizant of knowledge moving forward. For example, most in the radiation oncology community were unaware of insufficiency fractures in the sacrum associated with radiation therapy. However, with symmetric application of extended targets beyond the gross tumor volume defined by pre-sacral lymph nodes, if one is not careful full dose can be applied through the planning target volume (PTV) if the targets are applied in a symmetric manner. With modern MR sequences, we see the fractures on occasion at radiation doses under 6000 cGy, therefore it is important to place dose gradients across the sacrum and try not to place full dose across the entire structure. Investigators have found the pre-therapy exercise programs designed to maintain flexibility supports treatment reproducibility. This will become an important aspect of survivorship programs [64–72].

6. Breast

Patients with breast cancer will continue to require significant attention to detail by physics and dosimetry in the development of their care path for radiation therapy treatment planning. Because of irregular topography and multiple sloped surfaces of the breast which are patient specific, breast planning will not easily succumb to exclusive planning through AI. Optimizing radiation therapy treatment plans remains a balance of constraints. With modern treatment planning techniques, most breast geometries can be planned with no more than 5% of the breast target receiving 107% of the prescription dose when treatment is applied to breast tissue. Contouring, however, must be carefully done as breast requires both the radiation oncologist and the planning team to simultaneously think in 2, 3, and 4 dimensions to achieve objectives of management including contouring regional anatomy and establishing goals. If the oncologist contours from midline to the latissimus muscle in a left-sided breast patient, it is difficult to meet normal tissue constraints for the left ventricle, mean heart dose, and left pulmonary parenchyma as defined by most clinical trials. If arcs are applied, often constraints cannot be met for total lung and contralateral breast if tumor target dose homogeneity constraints are too strict. In these circumstances, the planning engine wants to please the planning team, therefore if constraints to normal tissue are not included in the planning strategy, the engines will work to limit radiation dose asymmetry through the target. However, the more homogeneous the plan becomes through the therapy target/breast, the more dose is "pushed" into normal tissue including heart/lung, and contralateral breast. Therefore, the planning teams and the radiation oncologist must provide balance to constraints to meet both tumor coverage without exaggerated dose to target volumes and limit dose to normal tissue. One of the key elements to planning breast radiation is to contour breast tissue without intentionally contouring volumes to midline or posteriorly to the latissimus. When possible, contour breast tissue including the surgical cavity in the target volume. Because of applying underlying cardiac and pulmonary constraints, dose will naturally be "pushed" to the medial and lateral chest wall parenchyma. If one contours to midline, dose can easily be pushed well into the contralateral breast without intent due to the attempt to provide full coverage to this point and well as the latissimus laterally. As the target becomes more inferior, therefore is less breast tissue in both the medial and lateral planes, providing an opportunity to titrate volumes in these locations and limit cardiac dose for left-sided patients. Therefore, contouring becomes a very important aspect to breast cancer radiation therapy. If the volumes are over-drawn or drawn in a casual manner, there is less physics and dosimetry can do to improve the situation. In a similar manner, strategies for regional care need to be carefully designed to balance dose to target and normal tissue. The internal mammary lymph nodes follow the anatomy of the internal mammary artery and there is direct drainage to level 3 and the sub-pectoral region as well as the axilla, therefore strategy must be applied to extent of regional coverage as seen in **Figure 6** as normal tissue dose is influenced by the contour and dose assigned to the target. Although anatomic guidelines have been established for clinical protocols, recent publications have suggested that regional failures can occur beyond the target volumes defined from anatomical guidelines, therefore we must identify high risk patients and possibly extend our field edges beyond our traditional boundaries. Patients with triple negative disease often now undergo positron emission tomography prior to initiating therapy. In these selected patients, nodal disease is at times identified beyond traditional field boundaries, therefore volume of regional disease now is driven by imaging.

Breast remains a disease requiring thoughtful application of radiation therapy to the breast, surgical cavity, and regional targets. Because outcomes remain excellent, attention to detail including normal tissue constraints is essential to modern planning [73–91].

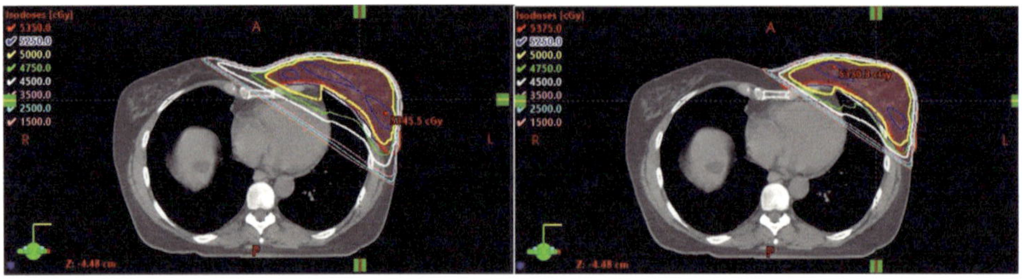

Figure 6.
MRT plan for breast radiotherapy with intentional (left) and unintentional (R) inclusion of internal mammary structures. Note difference in cardiac dose.

7. Extremities

For both adults and children, treatment of the extremities requires significant attention to detail. There are multiple tissue compartments that can be affected by therapy and each can have consequence for late effects of management. These can include major blood vessels, bone, joint spaces, nerve, and muscle bundles. Children require additional conformal avoidance to growth plates when possible for optimal outcome for growth and development. This is an area where natural barriers may be helpful in placing dose gradients across structures to limit late effects, especially bone and joint spaces.

Because the extremities resemble cylinders, volume modulated arcs often play an important role in radiation therapy planning. It is important to spare a strip of normal tissue as it extends through the cylinder. Although most oncologists try to generate the strip in a linear manner, is it likewise clinically acceptable to spare tissue in a non-linear serpentine manner if the strip remains non-interrupted. Of equal importance is the concept of immobilization. Historically, radiation oncologists approached this from a perspective that casts and rigid structures provided the optimal security for treatment reproducibility. However, even within a rigid structure, alterations in anatomy can occur either on a pre- or post-operative basis

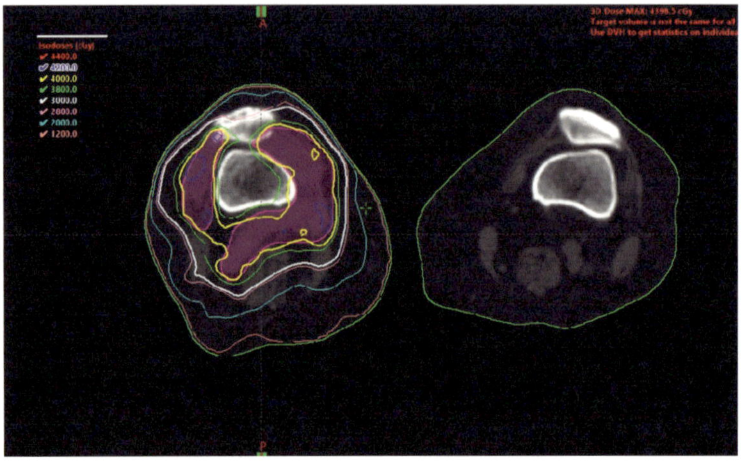

Figure 7.
Near circumferential extremity target with popliteal adenopathy. Note sparing of normal tissue in the posterior compartment.

DOI: http://dx.doi.org/10.5772/intechopen.105910

rendering the utility of rigid structures. Modern alignment and optical tracking tools have altered our approach to this situation. With the generation of these tools, we can build a "virtual cast" and maintain alignment without the rigor of rigid tools. Patients are more comfortable, we are more secure in daily treatment reproducibility, and treatment time is considerably decreased further contributing to security in daily set up. The volumes in these cases lend well to advanced technology therapy and conformal avoidance of key structures can be more readily accomplished. **Figure 7** represents a patient with an extremity lymphoma with popliteal adenopathy using volume modulated arc treatment to treat the circumferential target with sparing oof posterior soft tissue for lymphatic drainage [92–99].

8. Pediatrics

Pediatrics is a special subgroup of patients requiring considerable care and attention to detail. This population requires many departments within the hospital setting including anesthesia and social services to manage and optimize patient care. Often the targets treated in children are both similar and dissimilar to adult counterparts. CNS tumors often require similar normal tissue avoidance strategies with careful attention to the volume of temporal lobe treated including essential structures including but not limited to the chiasm, cochlea, retina, and brain stem. Diseases including germ cell tumors and medulloblastoma require therapy to the CNS fluid pathways, therefore strategy of care is different than adults and includes the craniospinal axis. Intensity modulation and protons have been used to limit exit dose to cardiac, pulmonary, liver, bowel, vertebral body, and renal volumes. Pediatric diseases often requiring whole lung therapy and strategies are now more routine to provide conformal avoidance of the heart using four-dimensional technology and volume modulated arc therapy. Neuroblastoma, sarcoma and other soft/bone tissue primary lesions often require advanced technology therapy application with a growing interest in theranostics for neuroblastoma care. In NCTN clinical trials, more than 25% of pediatric patients are now treated with protons when they receive radiation therapy, and this number is expected to increase moving forward. **Figure 8** is a 7-year-old patient

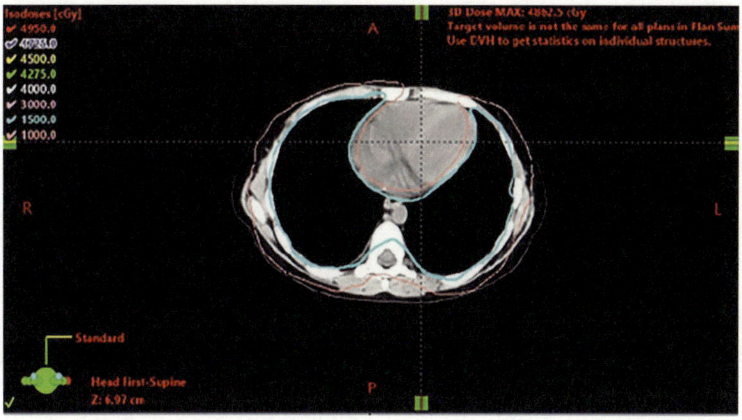

Figure 8.
Conformal avoidance of the heart in a 7-year-old patient during whole lung radiotherapy.

requiring whole lung therapy being treated with cardiac avoidance. Note the ability to limit dose to the cardiac structures using volume modulated arc therapy [93, 99–106].

9. Imaging in radiation oncology

With the advent of volumetric radiation therapy planning systems, imaging has become essential for modern therapy including dosimetry. Prior to the development of computer tomography-based simulation, patients were treated with two-dimensional planning with fluoroscopy simulators. Computer tomography simulation was a paradigm shift in radiation oncology. Today imaging is the infrastructure to all elements of activity in the department. Thoracic and upper abdominal patients are simulated with four-dimensional imaging. Many head and neck and nearly all CNS patients are planned with fusion imaging to accurately define target volumes of interest. Many CNS patients are now planned with multiple MR sequences which when applied are used for dose painting on clinical protocols using FLAIR, spectroscopy, and contrast as areas to target. Cone beam computer tomography has been incorporated into linear accelerators for daily target validation and can be applied for adaptive radiation therapy planning. Portal imagers have a dual role as a dosimeter. Single photon emission computed tomography (SPECT) imaging will play an essential role in computational analytics for theranostics at multiple time points to evaluate dose to volume and migration. Today, all patients are treated with image guidance to ensure stability and reproducibility in daily positioning. Optical tracking tools are used to ensure three-dimensional stability of patient set up during treatment and imaging is used to validate outcome. These changes in daily work and workflow have largely occurred during the past two decades, therefore physics staff and dosimetrists have become nimble in image acquisition and fusion. Modern planners have become expert in radiographic anatomy and often are responsible for contouring normal tissues including subtle structures such as the optic chiasm and the cochlea. These skills are far different than the skill set required two decades ago, and this evolving expertise is an important aspect to modern planning teams [1–11].

10. The role of AI in radiation oncology

As treatment planning becomes more complex, tasks that can be performed by AI will become invaluable to department workflow and an essential tool for planning teams. There are evolving standards in radiation oncology which must include the tools of artificial intelligence to optimize patient care moving forward. Departments will need to become as efficient as possible, even in performing tasks with advanced complexity. There are changing reimbursement models for radiation oncology and it is anticipated that compressed fractionation models will continue to mature. Therefore, even though the number of treatments to patient may decrease, there will be no decrease in the number and complexity of plans created for patients which will place planning teams including dosimetrists at the crossroad between devoting time to optimize a patient plan and efficiency of time applied to the task. We will need AI tools to help support and facilitate plan development. There are tools available for auto-contouring normal tissue and AI tools are in development to support planning including radiosurgery including the plan seen in **Figure 2**. This will save time and effort for planning teams and permit planners to focus on optimizing plans relative to constraints established for each patient [107].

11. Dosimetry in clinical trials

Clinical trials including the NCTN and industry are the primary vehicles used to validate process improvements in patient care. Quality assurance in clinical trials is important as the process ensures consistency in trial execution and we can trust the outcome of the trial. Physics and dosimetry teams play an important role in the development and execution of clinical trials. Members of the Imaging and Radiation Oncology Core (IROC) will write guidelines and data management strategies into each clinical trial for imaging and radiation oncology. IROC will review clinical trial objects in real time to ensure completeness of the data acquisition process and re-compute patient analytics for target and normal tissue dose in a single platform to harmonize dose to volume for each patient on study. It is only through processes such as this that data can be powered to understand toxicity relative to dose/volume and ensure protocol coverage of tumor targets. Moving forward, large databases housed in a uniform format with planning and outcome imaging will be required to optimize our understanding of tumor contouring and normal tissue metrics. Annotation of these cases can be used for the development of AI programs. Clinical trials remain our primary resource to confirm standards of care and the quality assurance physics and dosimetry teams of IROC serve an important role in the development and standardization of standards. The Cancer Imaging Archive (TCIA) will be an important resource to further enhance our knowledge by associating radiation therapy data with pathomics and radiomics [1–11].

12. Conclusions

Radiation therapy will play an important role for oncology care moving forward. With increasing evidence of the success treating patients with oligometastasis and refinement of tools for radiosurgery including protons, theranostics, and particle therapies, radiation oncology will remain an important component for patient care for decades to come in nearly all disease areas using sophisticated treatment strategies. An increasing number of patients with cancer are now treated with curative intent and the plans developed for these patients is becoming increasingly complex with varied normal tissue constraints, especially seen in patients requiring re-treatment. Many disease programs including breast and prostate are being treated with compressed fractionation, therefore there will be a decrease in the number of treatments, yet the number of plans will increase, and the complexity of each plan will increase. Therefore, the skill set of the planning teams will not resemble the skill set required for work a decade ago. Tomorrow's planning team will be fluent in AI for plan optimization and applied imaging/image fusion for target definition. Planning teams will require skill and knowledge to create sophisticated treatment plans in a timeframe required for timely care. Although an exciting new era for radiation oncology, the skill required for treatment planning and treatment execution is increasing and departments will need to provide educational resources to meet this need moving forward. Training programs will need to adapt to meet these evolving standards. The future of radiation oncology is strong; however, we must continue the process of self-improvement to meet the needs of patients moving forward.

Conflict of interest

The authors declare no conflict of interest.

Author details

Linda Ding[1], Carla D. Bradford[1], Kenneth Ulin[1], Koren Smith[1], I-Lin Kuo[1], Yankhua Fan[1], Abdulnasser Khalifeh[1], Fenghong Liu[1], Suhong Lu[1], Harry Bushe[1], Salvatore Larosa[1], Camelia Bunaciu[1], Jonathan Saleeby[1], Shannon Higgins[1], Julie Trifone[1], Maureen Britton[1], Joshua Taylor[1], Marious Croos[1], Katie Figura[1], Thomas Quinn[1], Linda O'Connor[1], Kathleen Briggs[1], Sherri Suhl[1], Jean Quigley[1], Heather Reifler[1], Shawn Kirby[1], Fred Prior[2], Joel Saltz[3], Maryann Bishop-Jodoin[1] and Thomas J. FitzGerald[1*]

1 Department of Radiation Oncology, UMass Chan Medical School, Worcester, MA, USA

2 Department of Biomedical Informatics, University of Arkansas for Medical Sciences, Little Rock, AR, USA

3 Department of Biomedical Informatics, Stony Brook School of Medicine, Stony Brook, NY, USA

*Address all correspondence to: tj.fitzgerald@umassmemorial.org

References

[1] FitzGerald TJ, Urie M, Ulin K, Laurie F, Yorty J, Hanusik R, et al. Processes for quality improvements in radiation oncology clinical trials. International Journal of Radiation Oncology—Biology—Physics. 2008;**71**(S1):S76-S79

[2] FitzGerald TJ, Followill DS, Galvin J, Knopp MV, Michalski JM, Rosen MA. Restructure of clinical trial system addresses evolving needs. ASTROnews. 2014 Winter;**17**(4):18-19

[3] Mendenhall NP, Meyer J, Williams J, Tebbi C, Kessel S, Laurie F, et al. The impact of central quality assurance prior to radiation therapy Hodgkin's disease. Blood. 2005;**106**(11):753

[4] FitzGerald TJ, Bishop-Jodoin M, Bosch WR, Curran WJ, Followill DS, Galvin JM, et al. Future vision for quality assurance of oncology clinical trials. Frontiers in Oncology. 2013;**3**:31

[5] Dharmarajan KV, Friedman DL, Schwartz CL, Chen L, FitzGerald TJ, McCarten KM, et al. Patterns of relapse from a phase 3 study of response-based therapy for intermediate-risk Hodgkin lymphoma (AHOD0031): A report from the Children's oncology group. International Journal of Radiation Oncology—Biology—Physics. 2015;**92**(1):60-66

[6] FitzGerald TJ, Bishop M, Followill DDS, Galvin J, Knopp MV, Michalski JM, et al. Imaging and data acquisition in clinical trials for radiation therapy. International Journal of Radiation Oncology—Biology—Physics. 2016;**94**(2):404-411

[7] FitzGerald TJ, Followill D, Laurie F, Boterberg T, Hanusik R, Kessel S, et al. Quality assurance in radiation oncology. Pediatric Blood & Cancer. 2021;**68**(Suppl 2):e38609. DOI: 10.1002/pbc.28609

[8] FitzGerald TJ, Bishop-Jodoin M, Cicchetti MG, Hanusik R, Kessel S, Laurie F, et al. Quality of radiotherapy reporting in randomized controlled trials of Hodgkin's lymphoma and non-Hodgkin's lymphoma. in regard to Bekelman and Yahalom (Int J Radiat Oncol Biol Phys 2009;73:492-498). International Journal of Radiation Oncology—Biology—Physics. 2010;**77**(1):315-316

[9] FitzGerald TJ. What we have learned: The impact of quality from a clinical trials perspective. Seminars in Radiation Oncology. 2012;**22**(1):18-28

[10] Gong YU, Yu J, Pang D, Zhen H, Galvin J, Xiao Y. Automated extraction of dose volume statistics for radiotherapy treatment plan evaluation in clinical trial quality assurance. Frontiers in Oncology. 2016;**6**:47. DOI: 10.3389/fonc.2016.00047

[11] Followill DS, Urie M, Galvin JM, Ulin K, Xiao Y, Fitzgerald TJ. Credentialing for participation in clinical trials. Frontiers in Oncology. 2012;**2**:198. DOI: 10.3389/fonc.2012.00198

[12] Wagner TH, Meeks SL, Bova FJ, Friedman WA, Willoughby TR, Kupelian PA, et al. Optical tracking technology in stereotactic radiation therapy. Medical Dosimetry. 2007 Summer;**32**(2):11-20

[13] Yock TI, Yeap BY, Ebb DH, Weyman E, Eaton BR, Sherry NA, et al. Long-term effects of proton therapy for pediatric medulloblastoma: A phase 2 single-arm study. The Lancet Oncology. 2016;**17**(3):287-298

[14] Indelicato DJ, Flampouri S, Rotondo RL, Bradley JA, Morris CG, Aldana PR, et al. Incidence and dosimetric parameters of pediatric brainstem toxicity following proton therapy. Acta Oncologica. 2014;**53**(10):1298-1304

[15] Roddy E, Mueller S. Late effects of treatment of pediatric central nervous system tumors. Journal of Child Neurology. 2016;**31**(2):237-254

[16] Nanda R, Dhabbaan A, Janss A, Shu HK, Esiashvili N. The feasibility of frameless stereotactic radiosurgery in the management of pediatric central nervous system tumors. Journal of Neuro-Oncology. 2014;**117**(2):329-335

[17] Rieber JG, Kessel KA, Witt O, Behnisch W, Kulozik AE, Debus J, et al. Treatment tolerance of particle therapy in pediatric patients. Acta Oncologica. 2015;**54**(7):1049-1055

[18] Freund D, Zhang R, Sanders M, Newhauser W. Predictive risk of radiation induced cerebral necrosis in pediatric brain cancer patients after VMAT versus proton therapy. Cancers (Basel). 2015;**7**(2):617-630

[19] Giraud P, Kantor G, Loiseau H, Rosenzweig KE. Target definition in the thorax and central nervous system. Seminars in Radiation Oncology. 2005;**15**(3):146-156

[20] Kim TG, Lim DH. Interfractional variation of radiation target and adaptive radiotherapy for totally resected glioblastoma. Journal of Korean Medical Science. 2013;**28**(8):1233-1237

[21] Dilalla V, Chaput G, Williams T, Sultanem K. Radiotherapy side effects: Integrating a survivorship clinical lens to better serve patients. Current Oncology. 2020;**27**(2):107-112

[22] Mendel JT, Jaster AW, Yu FF, Morris LC 3rd, Lynch PT, Shah BR, et al. Fundamentals of radiation oncology for neurologic imaging. Radiographics. 2020;**40**(3):827-858

[23] Castellano A, Bailo M, Cicone F, Carideo L, Quartuccio N, Mortini P, et al. Advanced imaging techniques for radiotherapy planning of gliomas. Cancers (Basel). 2021;**13**(5):1063

[24] Sonke JJ, Aznar M, Rasch C. Adaptive radiotherapy for anatomical changes. Seminars in Radiation Oncology. 2019;**29**(3):245-257

[25] Chen AM, Daly ME, Cui J, Mathai M, Benedict S, Purdy JA. Clinical outcomes among patients with head and neck cancer treated by intensity-modulated radiotherapy with and without adaptive replanning. Head & Neck. 2014;**36**(11):1541-1546

[26] Hou J, Guerrero M, Suntharalingam M, D'Souza WD. Response assessment in locally advanced head and neck cancer based on RECIST and volume measurements using cone beam CT images. Technology in Cancer Research & Treatment. 2015;**14**(1):19-27

[27] Noble DJ, Yeap PL, Seah SYK, Harrison K, Shelley LEA, Romanchikova M, et al. Anatomical change during radiotherapy for head and neck cancer, and its effect on delivered dose to the spinal cord. Radiotherapy and Oncology. 2019;**130**:32-38

[28] Bertholet J, Anastasi G, Noble D, Bel A, van Leeuwen R, Roggen T. Patterns of practice for adaptive and real-time radiation therapy (POP-ART RT) part II: Offline and online plan adaption for interfractional changes. Radiotherapy and Oncology. 2020;**153**:88-96

[29] Beltran M, Ramos M, Rovira JJ, Perez-Hoyos S, Sancho M, Puertas E, et al.

Dose variations in tumor volumes and organs at risk during IMRT for head-and-neck cancer. Journal of Applied Clinical Medical Physics. 2012;**13**(6):101-111

[30] Contesini M, Guberti M, Saccani R, Braglia L, Iotti C, Botti A, et al. Setup errors in patients with head-neck cancer (HNC), treated using the intensity modulated radiation therapy (IMRT) technique: How it influences the customised immobilisation systems, patient's pain and anxiety. Radiation Oncology. 2017;**12**(1):72

[31] National Institutes of Health. Surveillance, Epidemiology, End Results. Cancer Stat Facts. 2022. Available from: https://seer.cancer.gov/statfacts/html/all.html [Accessed: Jun 13, 2022]

[32] Elhalawani H, Elgohari B, Lin TA, Mohamed ASR, Fitzgerald TJ, Laurie F, et al. An in-silico quality assurance study of contouring target volumes in thoracic tumors within a cooperative group setting. Clinical and Translational Radiation Oncology. 2019;**15**:83-92. Erratum in: Clin Transl Radiat Oncol. 2021;27:96-7

[33] Schimek-Jasch T, Troost EG, Rücker G, Prokic V, Avlar M, Duncker-Rohr V, et al. A teaching intervention in a contouring dummy run improved target volume delineation in locally advanced non-small cell lung cancer: Reducing the interobserver variability in multicentre clinical studies. Strahlentherapie und Onkologie. 2015;**191**(6):525-533

[34] Giraud P, Elles S, Helfre S, De Rycke Y, Servois V, Carette MF, et al. Conformal radiotherapy for lung cancer: Different delineation of the gross tumor volume (GTV) by radiologists and radiation oncologists. Radiotherapy and Oncology. 2002;**62**(1):27-36

[35] Spoelstra FO, Senan S, Le Péchoux C, Ishikura S, Casas F, Ball D, et al. Variations in target volume definition for postoperative radiotherapy in stage III non-small-cell lung cancer: Analysis of an international contouring study. International Journal of Radiation Oncology, Biology, Physics. 2010;**76**(4):1106-1113

[36] Hanna GG, Carson KJ, Lynch T, McAleese J, Cosgrove VP, Eakin RL, et al. 18F-fluorodeoxyglucose positron emission tomography/computed tomography-based radiotherapy target volume definition in non-small-cell lung cancer: Delineation by radiation oncologists vs. joint outlining with a PET radiologist? International Journal of Radiation Oncology, Biology, Physics. 2010;**78**(4):1040-1051

[37] Greco C, Rosenzweig K, Cascini GL, Tamburrini O. Current status of PET/CT for tumour volume definition in radiotherapy treatment planning for non-small cell lung cancer (NSCLC). Lung Cancer. 2007;**57**(2):125-134

[38] Jouin A, Pourel N. Target volume margins for lung cancer: internal target volume/clinical target volume. Cancer Radiothérapie. 2013;**17**(5-6):428-433

[39] Konert T, Vogel WV, Paez D, Polo A, Fidarova E, Carvalho H, et al. Introducing FDG PET/CT-guided chemoradiotherapy for stage III NSCLC in low- and middle-income countries: Preliminary results from the IAEA PERTAIN trial. European Journal of Nuclear Medicine and Molecular Imaging. 2019;**46**(11):2235-2243

[40] Kumar S, Holloway L, Roach D, Pogson E, Veera J, Batumalai V, et al. The impact of a radiologist-led workshop on MRI target volume delineation for radiotherapy. Journal of Medical Radiation Sciences. 2018;**65**(4):300-310

[41] Rooney MK, Zhu F, Gillespie EF, Gunther JR, McKillip RP,

Lineberry M, et al. Simulation as more than a treatment-planning tool: A systematic review of the literature on radiation oncology simulation-based medical education. International Journal of Radiation Oncology, Biology, Physics. 2018;**102**(2):257-283

[42] Roy AE, Wells P. Volume definition in radiotherapy planning for lung cancer: How the radiologist can help. Cancer Imaging. 2006;**6**(1):116-123

[43] Dobbs HJ. Defining the radiation target on a daily basis. Cancer Imaging. 2006;**6**(1):30-32

[44] Jin JY, Ajlouni M, Chen Q, Yin FF, Movsas B. A technique of using gated-CT images to determine internal target volume (ITV) for fractionated stereotactic lung radiotherapy. Radiotherapy and Oncology. 2006;**78**(2):177-184

[45] Lefkopoulos D, Ferreira I, Isambert A, Le Péchoux C, Mornex F. Present and future of the image guided radiotherapy (IGRT) and its applications in lung cancer treatment. Cancer Radiothérapie. 2007;**11**(1-2):23-31

[46] Mageras GS, Mechalakos J. Planning in the IGRT context: Closing the loop. Seminars in Radiation Oncology. 2007;**17**(4):268-277

[47] Fay M, Poole CM, Pratt G. Recent advances in radiotherapy for thoracic tumors. Journal of Thoracic Disease. 2013;**5**(Suppl 5):S551-S555

[48] Yang Y, Zhou S, Shang P, Qi E, Wu S, Xie Y. Contour propagation using feature-based deformable registration for lung cancer. BioMed Research International. 2013;**2013**:701514. DOI: 10.1155/2013/701514

[49] Wu G, Lian J, Shen D. Improving image-guided radiation therapy of lung cancer by reconstructing 4D-CT from a single free-breathing 3D-CT on the treatment day. Medical Physics. 2012;**39**(12):7694-7709

[50] Peng T, Wang Y, Xu TC, Shi L, Jiang J, Zhu S. Detection of lung contour with closed principal curve and machine learning. Journal of Digital Imaging. 2018;**31**(4):520-533

[51] Hong JC, Salama JK. Dose escalation for unresectable locally advanced non-small cell lung cancer: End of the line? Translational Lung Cancer Research. 2016;**5**(1):126-133

[52] Desideri I, Loi M, Francolini G, Becherini C, Livi L, Bonomo P. Application of radiomics for the prediction of radiation-induced toxicity in the IMRT era: Current state-of-the-art. Frontiers in Oncology. 2020;**10**:1708. DOI: 10.3389/fonc.2020.01708

[53] Aerts HJ, Velazquez ER, Leijenaar RT, Parmar C, Grossmann P, Carvalho S, et al. Decoding tumor phenotype by noninvasive imaging using a quantitative radiomics approach. Nature Communications. 2014;**5**:4006. DOI: 10.1038/ncomms5006. Erratum in: Nature Communications 2014;5:4644

[54] Yakar M, Etiz D. Artificial intelligence in radiation oncology. Artificial Intelligence in Medical Imaging. 2021;**2**(2):13-31

[55] Meyer P, Noblet V, Mazzara C, Lallement A. Survey on deep learning for radiotherapy. Computers in Biology and Medicine. 2018;**98**:126-146

[56] Lecun Y, Bengio Y, Hinton G. Deep learning. Nature. 2015;**521**(7553):436-444

[57] Lustberg T, van Soest J, Gooding M, Peressutti D, Aljabar P, van der Stoep J, et al. Clinical evaluation of atlas and deep

learning based automatic contouring for lung cancer. Radiotherapy and Oncology. 2018;**126**(2):312-317

[58] Thor M, Apte A, Haq R, Iyer A, LoCastro E, Deasy JO. Using auto-segmentation to reduce contouring and dose inconsistency in clinical trials: The simulated impact on RTOG 0617. International Journal of Radiation Oncology, Biology, Physics. 2021;**109**(5):1619-1626

[59] Grossmann P, Stringfield O, El-Hachem N, Bui MM, Rios Velazquez E, Parmar C, et al. Defining the biological basis of radiomic phenotypes in lung cancer. eLife. 2017;**6**:e23421. DOI: 10.7554/eLife.23421

[60] Jarrett D, Stride E, Vallis K, Gooding MJ. Applications and limitations of machine learning in radiation oncology. The British Journal of Radiology. 2019;**92**(1100):20190001. DOI: 10.1259/bjr.20190001

[61] Deig CR, Kanwar A, Thompson RF. Artificial intelligence in radiation oncology. Hematology/Oncology Clinics of North America. 2019;**33**(6):1095-1104

[62] Cherezov D, Goldgof D, Hall L, Gillies R, Schabath M, Müller H, et al. Revealing tumor habitats from texture heterogeneity analysis for classification of lung cancer malignancy and aggressiveness. Scientific Reports. 2019;**9**(1):4500

[63] Gillies RJ, Kinahan PE, Hricak H. Radiomics: Images are more than pictures, they are data. Radiology. 2016;**278**(2):563-577

[64] Rudra S, Jiang N, Rosenberg SA, Olsen JR, Roach MC, Wan L, et al. Using adaptive magnetic resonance image-guided radiation therapy for treatment of inoperable pancreatic cancer. Cancer Medicine. 2019;**8**(5):2123-2132

[65] Hall WA, Paulson ES, van der, Heide UA, Fuller CD, Raaymakers BW, Lagendijk JJW, et al. The transformation of radiation oncology using real time magnetic resonance guidance: A review. European Journal of Cancer. 2019;**122**:42-52

[66] Witt JS, Rosenberg SA, Bassetti MF. MRI-guided adaptive radiotherapy for liver tumours: Visualising the future. The Lancet Oncology. 2020;**21**(2):e74-e82. DOI: 10.1016/S1470-2045(20)30034-6

[67] Choi SH, Seong J. Strategic application of radiotherapy for hepatocellular carcinoma. Clinical and Molecular Hepatology. 2018;**24**(2):114-134

[68] Rosenberg SA, Henke LE, Shaverdian N, Mittauer K, Wojcieszynski AP, Hullett CR, et al. A multi-institutional experience of MR-guided liver stereotactic body radiation therapy. Advances in Radiation Oncology. 2018;**4**(1):142-149

[69] Boldrini L, Corradini S, Gani C, Henke L, Hosni A, Romano A, et al. MR-guided radiotherapy for liver malignancies. Frontiers in Oncology. 2021;**11**:616027. DOI: 10.3389/fonc.2021.616027

[70] Tetar SU, Bruynzeel AME, Oei SS, Senan S, Fraikin T, Slotman BJ, et al. Magnetic resonance-guided stereotactic radiotherapy for localized prostate cancer: Final results on patient-reported outcomes of a prospective phase 2 study. European Urology Oncology. 2021;**4**(4):628-634

[71] Maturen KE, Feng MU, Wasnik AP, Azar SF, Appelman HD, Francis IR, et al. Imaging effects of radiation therapy in the abdomen and pelvis: Evaluating "innocent bystander" tissues. Radiographics. 2013;**33**(2):599-619

[72] Sawicki M. Treatment planning in brachytherapy HDR based on three-dimensional image. In: Halefoglu AM, Computed Tomography—Advanced Applications. London: IntechOpen; 2017

[73] Blichert-Toft M, Rose C, Andersen JA, Overgaard M, Axelsson CK, Andersen KW, et al. Danish randomized trial comparing breast conservation therapy with mastectomy: Six years of life- table analysis. Danish breast cancer cooperative group. Journal of the National Cancer Institute. Monographs. 1992;**11**(11):19-25

[74] Borm KJ, Voppichler J, Düsberg M, Oechsner M, Vag T, Weber W, et al. FDG/PET CT-based lymph node atlas in breast cancer patients. International Journal of Radiation Oncology, Biology, Physics. 2019;**103**(3):574-582

[75] Borm KJ, Kessel K, Devecka M, Muench S, Straube C, Schiller K, et al. Variability in lymph node irradiation in patients with breast cancer-results from a multi-center survey in German-speaking countries. Strahlentherapie und Onkologie. 2020;**196**(1):15-22

[76] Citron ML, Berry DA, Cirrincione C, Hudis C, Winer EP, Gradishar WJ, et al. Randomized trial of dose-dense versus conventionally scheduled and sequential versus concurrent combination chemotherapy as postoperative adjuvant treatment of node-positive primary breast cancer: First report of intergroup trial C9741/cancer and leukemia group B trial 9741. Journal of Clinical Oncology. 2003;**21**(8):1431-1439

[77] Donker M, van Tienhoven G, Straver ME, Meijnen P, van de, Velde CJ, Mansel RE, et al. Radiotherapy or surgery of the axilla after a positive sentinel node in breast cancer (EORTC 10981-22023 AMAROS): A randomised, multicentre, open-label, phase 3 non-inferiority trial. The Lancet Oncology. 2014;**15**(12):1303-1310

[78] Emami B, Lyman J, Brown A, Coia L, Goitein M, Munzenrider JE, et al. Tolerance of normal tissue of therapeutic irradiation. International Journal of Radiation Oncology, Biology, Physics. 1991;**21**(1):109-122

[79] Fisher B, Anderson S, Bryant J, Margolese RG, Deutsch M, Fisher ER, et al. Twenty-year follow-up of a randomized trial comparing total mastectomy, lumpectomy, and lumpectomy plus irradiation for the treatment of invasive breast cancer. The New England Journal of Medicine. 2002;**347**(16):1233-1241

[80] Gentile MS, Usman AA, Neuschler EI, Sathiaseelan V, Hayes JP, Small W Jr. Contouring guidelines for the axillary lymph nodes for the delivery of radiation therapy in breast cancer: Evaluation of the RTOG breast cancer atlas. International Journal of Radiation Oncology, Biology, Physics. 2015;**93**(2):257-265

[81] Giuliano AE, Ballman KV, McCall L, Beitsch PD, Brennan MB, Kelemen PR, et al. Effect of axillary dissection vs no axillary dissection on 10-year overall survival among women with invasive breast cancer and sentinel node metastasis: The ACOSOG Z0011 (Alliance) randomized clinical trial. Journal of the American Medical Association. 2017;**318**(10):918-926

[82] Henderson IC, Berry DA, Demetri GD, Cirrincione CT, Goldstein LJ, Martino S, et al. Improved outcomes from adding sequential paclitaxel but not from escalating doxorubicin dose in an adjuvant chemotherapy regimen for patients with node-positive primary breast cancer. Journal of Clinical Oncology. 2003;**21**(6):976-983

[83] Jacobson JA, Danforth DN, Cowan KH, d'Angelo T, Steinberg SM, Pierce L, et al. Ten-year results of a comparison of conservation with mastectomy in the treatment of stage I and II breast cancer. The New England Journal of Medicine. 1995;**332**(14):907-911

[84] Jagsi R, Chadha M, Moni J, Ballman K, Laurie F, Buchholz TA, et al. Radiation field design in the ACOSOG Z0011 (Alliance) trial. Journal of Clinical Oncology. 2014;**32**(32):3600-3606

[85] Morris AD, Morris RD, Wilson JF, White J, Steinberg S, Okunieff P, et al. Breast-conserving therapy vs mastectomy in early stage breast cancer: A meta-analysis of 10-year survival. The Cancer Journal from Scientific American. 1997;**3**(1):6-12

[86] Reznik J, Cicchetti MG, Degaspe B, Fitzgerald TJ. Analysis of axillary coverage during tangential radiation therapy to the breast. International Journal of Radiation Oncology, Biology, Physics. 2005;**61**(1):163-168

[87] Sarrazin D, Lê MG, Arriagada R, Contesso G, Fontaine F, Spielmann M, et al. Ten-year results of a randomized trial comparing a conservative treatment to mastectomy in early breast cancer. Radiotherapy and Oncology. 1989;**14**(3):177-184

[88] Sartor CI, Peterson BL, Woolf S, Fitzgerald TJ, Laurie F, Turrisi AJ, et al. Effect of addition of adjuvant paclitaxel on radiotherapy delivery and locoregional control of node-positive breast cancer: Cancer and leukemia group B 9344. Journal of Clinical Oncology. 2005;**23**(1):30-40

[89] Siddon RL, Buck BA, Harris JR, Svensson GK. Three-field technique for breast irradiation using tangential field corner blocks. International Journal of Radiation Oncology, Biology, Physics. 1983;**9**(4):583-588

[90] van, Dongen JA, Voogd AC, Fentiman IS, Legrand C, Sylvester RJ, Tong D, et al. Long-term results of a randomized trial comparing breast-conserving therapy with mastectomy: European Organization for Research and Treatment of cancer 10801 trial. Journal of the National Cancer Institute. 2000;**92**(14):1143-1150

[91] Veronesi U, Saccozzi R, Del Vecchio M, Banfi A, Clemente C, De Lena M, et al. Comparing radical mastectomy with quadrantectomy, axillary dissection, and radiotherapy in patients with small cancers of the breast. The New England Journal of Medicine. 1981;**305**(1):6-11

[92] Cho CJ, Catton C, Holloway CL, Goddard K. Patterns of practice survey: radiotherapy for soft tissue sarcoma of the extremities. Cureus. 2019;**11**(11):e6153. DOI: 10.7759/cureus.6153

[93] Zheng X, Dai T, Shu X, Pu Y, Feng G, Li X, et al. A new method of lower extremity immobilization in radiotherapy. Radiation Oncology. 2012;7:27

[94] Tiwana MS, Barnes M, Kiraly A, Olson RA. Utilization of palliative radiotherapy for bone metastases near end of life in a population-based cohort. BMC Palliative Care. 2016;**15**:2

[95] Christiansen SA, Cohen P, Broome C. A case series of non-Hodgkin's lymphoma recurrence in lower extremities mimicking deep vein thrombosis. Blood. 2016;**128**(22):5344

[96] Dargent JL, Lespagnard L, Feoli F, Debusscher L, Greuse M, Bron D. De novo CD5-positive diffuse large B-cell

lymphoma of the skin arising in chronic limb lymphedema. Leukemia & Lymphoma. 2005;**46**(5):775-780

[97] González-Vela MC, González-López MA, Val-Bernal JF, Fernández-Llaca H. Cutaneous diffuse large B-cell lymphoma of the leg associated with chronic lymphedema. International Journal of Dermatology. 2008;**47**(2):174-177

[98] Ishii N, Aoki M, Kishi K. Lymphedema of the lower extremities due to refractory malignant lymphoma treated by lymphaticovenous anastomosis. Plastic and Reconstructive Surgery. Global Open. 2017;**5**(8):e1475. DOI: 10.1097/GOX.0000000000001475

[99] Persky DO, Unger JM, Spier CM, Stea B, LeBlanc M, McCarty MJ, et al. Phase II study of rituximab plus three cycles of CHOP and involved-field radiotherapy for patients with limited-stage aggressive B-cell lymphoma: Southwest oncology group study 0014. Journal of Clinical Oncology. 2008;**26**(14):2258-2263

[100] Kalapurakal JA, Gopalakrishnan M, Walterhouse DO, Rigsby CK, Rademaker A, Helenowski I, et al. Cardiac sparing whole lung IMRT in patients with pediatric tumors and lung metastasis: Final report of a prospective multicenter trial. International Journal of Radiation Oncology, Biology, Physics. 2019;**103**(1):28-37

[101] Kalapurakal JA, Zhang Y, Kepka A, Zawislak B, Sathiaseelan V, Rigsby C, et al. Cardiac sparing whole lung IMRT in children with lung metastasis. International Journal of Radiation Oncology, Biology, Physics. 2013;**85**(3):761-767

[102] Kalapurakal JA, Lee B, Bautista J, Rigsby C, Helenowski I, Gopalakrishnan M. Cardiac-sparing whole lung intensity modulated radiation therapy in children with Wilms tumor: Final report on technique and abdominal field matching to maximize normal tissue protection. Practical Radiation Oncology. 2019;**9**(1):e62-e73. DOI: 10.1016/j.prro.2018.07.005

[103] Zhang X, Zhao KL, Guerrero TM, McGuire SE, Yaremko B, Komaki R, et al. Four-dimensional computed tomography-based treatment planning for intensity modulated radiation therapy and proton therapy for distal esophageal cancer. International Journal of Radiation Oncology, Biology, Physics. 2008;**72**(1):278-287

[104] Kole TP, Aghayere O, Kwah J, Yorke ED, Goodman KA. Comparison of heart and coronary artery doses associated with intensity-modulated radiotherapy versus three-dimensional conformal radiotherapy for distal esophageal cancer. International Journal of Radiation Oncology, Biology, Physics. 2012;**83**(5):1580-1586

[105] Scobioala S, Eich HT. Risk stratification of pulmonary toxicities in the combination of whole lung irradiation and high-dose chemotherapy for Ewing sarcoma patients with lung metastases: A review. Strahlentherapie und Onkologie. 2020;**196**(6):495-504

[106] Suzuki G, Ogata T, Aibe N, Yamazaki H, Yagyu S, Iehara T, et al. Effective heart sparing whole lung irradiation using volumetric arc therapy: A case report. Journal of Medical Case Reports. 2019;**13**(1):277

[107] Huynh E, Hosny A, Guthier C, Bitterman DS, Petit SF, Haas-Kogan DA, et al. Artificial intelligence in radiation oncology. Nature Reviews. Clinical Oncology. 2020;**17**(12):771-781

Chapter 2

Treatment of Head and Neck Cancers Using Radiotherapy

Wan Shun Leung and Hing Ming Hung

Abstract

Radiotherapy is one of the major treatments for head and neck cancers. This chapter discusses the importance of radiotherapy in treating the common types of head and neck cancers, which can be used as a primary treatment or as a postoperative adjuvant treatment to increase the survival of head and neck cancer patients. Because head and neck cancers are likely to be closely surrounded by radiation-sensitive vital organs, the dosimetric superiority of intensity-modulated radiotherapy (IMRT) to achieve highly conformal dose to the planning target volume (PTV) and avoidance of organs at risk (OARs) helps maintain the cornerstone role of radiotherapy in treating the disease. The rationale of IMRT and the treatment planning technique are introduced. Treatment planning of radiotherapy is one of the key procedures in IMRT. The inverse planning process involves many decision-making steps, including PTV and OAR delineation, beam arrangement settings, objective function setting, etc. These important steps are all illustrated in the chapter, with a specific discussion of planning challenges relevant to head and neck cancers. Finally, the promises for further development of IMRT in terms of OARs dose sparing and PTV dose escalation are briefly discussed and reviewed.

Keywords: radiotherapy, treatment planning, head and neck cancers, IMRT, VMAT

1. Introduction

This chapter aims to provide background information about head and neck cancers, including their respective treatment options and radiotherapy techniques. It is divided into 4 parts. Part 1 summarizes the information about head and neck cancers and the use of radiotherapy for head and neck cancers. Part 2 introduces the intensity-modulated radiotherapy (IMRT) which is commonly used in the treatment of head and neck cancers. Part 3 reviews the planning techniques of IMRT. Finally, part 4 discusses the current challenges of head and neck cancers radiotherapy and the promises to overcome the challenges.

2. Head and neck cancers

2.1 Epidemiology statistics

Head and neck cancers refer to the carcinomas that originate from any parts of the upper aero-digestive tract. They also include the cancers of the thyroid and salivary glands. Although head and neck cancers no longer rank among the top 5 cancers in the latest report [1], they are still regarded as major types of cancer in Hong Kong [2]. One of the main reasons for this recognition is that nasopharyngeal cancer (NPC) is ranked sixth in terms of the number of new cases in the male population in Hong Kong [1]. The NPC worldwide figures illustrated by the age-standardized rate (ASR) was 1.2 per 100,000 [3], which were much lower than the incidence in Hong Kong which was 7.4 per 100,000 in the year 2012 [1]. The high incidence of NPC in Hong Kong is attributed to its special geographical epidemiology pattern that 76% of new cases were found in east and south-eastern parts of Asia, in which Hong Kong is situated [4]. Other head and neck cancers recorded in the Hong Kong Cancer Registry include cancers of the lip, oral cavity, pharynx, nasal cavity, middle ear and accessory sinuses, larynx, and thyroid gland. Altogether, there were 2617 new cases of head and neck cancers reported in 2016 in Hong Kong, which accounted for 8.3% of all cancer new cases [1]. NPC was the most common type of head and neck cancer, accounting for 46.6% of all new cases. It was followed by the cancer of the tongue and larynx which accounted for 13.9% and 11.4%, respectively [1]. Although there have been some variations in the trend of ASR between sub-sites, the overall ASR of head and neck cancers in Hong Kong has remained around 21 per 100,000 in the past decade. Because of the relatively high incidence of head and neck cancers, their treatment remains one of the major burdens in the health care services in Hong Kong [2].

2.2 The role of radiotherapy in major types of head and neck cancers

The role of radiotherapy in the radical treatment of five types of head and neck cancers including cancers of the nasopharynx, oral cavity, larynx, maxillary sinus, and parotid gland is discussed in this section. Intensity-modulated radiotherapy is a standard radiotherapy technique used. The benefit of IMRT is that it is capable of delivering highly conformal doses to the target while sparing the nearby organs at risk (OARs).

2.2.1 Nasopharynx

Radiotherapy is the major treatment modality for nasopharyngeal carcinoma (NPC). It is because the primary tumor site of NPC is difficult to be accessed by surgical intervention, and the tumor cells of NPC are sensitive to radiation [5]. The use of radiotherapy alone is effective to treat stage I to II NPC, while concurrent chemotherapy is added for higher stages disease to achieve better local-regional control and survival outcome [6]. IMRT is the preferred radiotherapy technique and the late side effect of xerostomia in patients receiving IMRT was significantly reduced [7]. The current standard of the prescribed total dose to the primary tumor is to give 70 Gy in 33–35 fractions [8]. With the use of simultaneous integrated boost, the prophylactic dose which is lower than the dose to the primary tumor is prescribed for the potential microscopic spread of the primary tumor and selected cervical lymph nodes regions. The prophylactic prescription can be varied in different local practices, it was reported that the prescriptions for the intermediate and low-risk cervical lymph nodes were about 60 Gy and 50 Gy, respectively [8, 9].

2.2.2 Oral cavity

The cancer of the oral cavity includes various sub-sites such as the anterior tongue, buccal mucosa, hard palate, soft palate, alveolus, and floor of the mouth. The primary treatment of the cancer of the oral cavity varied according to the stage, which can be briefly divided into early and advanced. For early-stage which refers to T1 and early T2 tumors, radiotherapy entirely or partly delivered by brachytherapy can result in similar local control as in surgery [10, 11]. However, a recent retrospective study reported that primary radiotherapy to early-stage oral cavity cancer patients resulted in higher mortality as compared with those who received primary surgery [12]. It has also been reported in the same article that the majority (more than 95%) of early-stage oral cavity cancer patients received primary surgery. The small proportion of patients receiving primary radiotherapy in this group of patients was attributed to the fact that brachytherapy services were not available due to lack of expertise and suitability of applicator for insertion [10]. Hence, most early-stage oral cavity cancer patients receive surgery for primary treatment, although radiotherapy is also an alternative. Postoperative radiotherapy is only indicated for positive or close margins after resection [13]. For advanced oral cavity cancer, surgery is often the standard primary treatment whenever resectable [14], and then followed by adjuvant radiotherapy or chemo-radiotherapy. For non-resectable advanced oral cavity cancer, radical radiotherapy is offered in conjunction with chemotherapy or targeted therapy to improve disease control [15]. The total prescribed dose is 70 Gy to the gross tumor or 66 Gy to the tumor bed after resection, delivered with 2 Gy per fraction. Similar to NPC, prophylactic irradiation to the cervical lymph nodes regions is also used, where 60 Gy and 54 Gy are prescribed to the intermediate-risk and low-risk regions, respectively [16].

2.2.3 Larynx

A specific consideration when treating cancer of the larynx is preserving organs and function. Radiotherapy alone or concurrent chemoradiotherapy is the most widely applied approach in organ preservation therapy [17]. Radical surgery is the rival choice for the patients, the outcome would lead to sub-optimal quality of life because it would result in loss of voice, swallowing problem, and often a permanent tracheostomy. To achieve a better quality of life after treatment, organ preservation therapy using radiotherapy or chemoradiotherapy is recommended for early-stage disease and some advanced cases of T3 and T4 [17, 18]. The consideration of offering surgery instead of radical chemoradiotherapy for advanced cases includes patients' condition and the extent of the disease and should be assessed by an expert panel of clinicians from different disciplines [19, 20]. Even when surgery is chosen as the treatment option, radiotherapy still has the role in providing postoperative adjuvant treatment for high-grade tumors, positive margins, cervical lymph nodes involvement, and tumor invasion beyond the larynx [21]. The prescribed dose ranged from 66 Gy to 76 Gy to the primary tumor site and involved lymph node, and the prescription for the selective lymph node with suspected microscopic involvement is at least 50 Gy [22].

2.2.4 Maxillary sinus

Although the primary treatment of the cancer of the maxillary sinus is surgery, postoperative radiotherapy is indicated for stage 2 and stage 3 disease, and for stage 1 disease when the surgical margin is insufficient [20]. For locally advanced disease, induction chemotherapy and then concurrent chemoradiotherapy have been suggested for non-resectable patients [23]. The treatment outcome for these patients would be better if the tumor can be down-staged and subsequent resection is possible [23]. The concern of the radiotherapy to the maxillary sinus includes the preservation of the optic apparatus which are near to the tumor [20]. It has been reported that 37% of the patients who received conventional radiotherapy developed radiotherapy-induced blindness [24]. IMRT is the preferred technique. It has been reported that IMRT could significantly spare nearby organs than those in 3DCRT. The dose to the optic chiasm can be significantly reduced from over 60 Gy in 3DCRT to less than 40 Gy in IMRT [25], while the tumor coverage by the prescribed dose is increased from 83% in 3DCRT to 95% in IMRT. The prescribed dose to the primary tumor site ranged from 66 to 70 Gy.

2.2.5 Parotid gland

The primary treatment for the cancer of parotid gland is surgical resection. Radiotherapy is used for adjuvant postoperative treatment except in small and low histological risk tumor with clear surgical margins [26]. In addition, radiotherapy is also indicated as radical treatment in advanced parotid gland cancer cases when resection of the tumor is not possible [27]. The prescribed dose to the primary site is about 66 Gy. IMRT is advocated as the treatment technique to improve OARs sparing [28].

3. Intensity-modulated radiotherapy

As discussed, IMRT has commonly used for radiotherapy of head and neck cancers The concept of IMRT has been introduced as early as 30 years ago [29], when the method of optimizing the intensity distribution of the incident beams with the purpose to achieve the required dose distribution in the targets was described. The following points summarize the concept of the delivery of IMRT: (1) There are multiple radiation beams with specially decided nonuniform intensity in beamlets, also known as intensity modulation. (2) The multiple radiation beams are applied from different directions, and the region of the convergence of the beams can achieve the desired dose distribution based on the modulated beam intensity. (3) Calculation of the modulated beam intensity usually follows an inverse approach, in which the final dose distribution indicated by planners is used by the computer to calculate the intensity of each beamlets in the treatment field of the IMRT plan.

The delivery of intensity-modulated beams is largely contributed by the dynamic multi-leaf collimator (MLC). The MLC can change the field shape automatically and the summation of numerous sub-fields in different shapes then generate a field with intensity modulation. A simplified rationale of intensity modulation is illustrated in **Figure 1**. Assume there is no OAR surrounding the target, the intensity of the beam should be proportional to the target thickness from the perspective of each beam. Although beam modifying devices such as wedges and compensators have been used in 3DCRT, their flexibility of beam intensity modification is far less than that in the

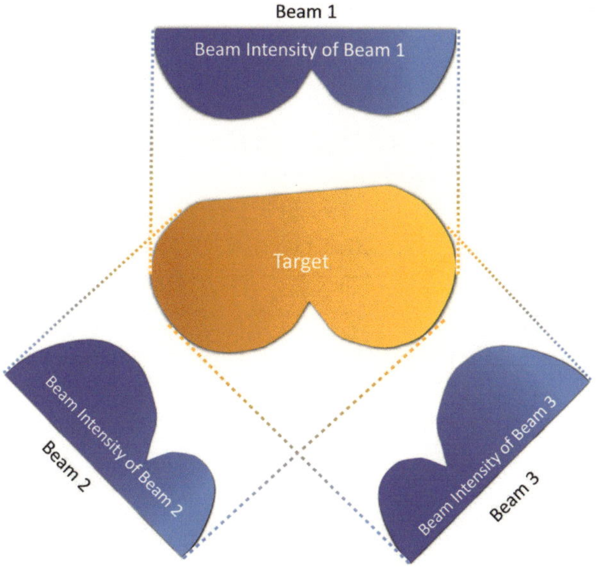

Figure 1.
Illustration of the relationship of beam intensity and target thickness.

IMRT. This is best illustrated by the fact that IMRT can produce concave shape isodose distribution which 3DCRT can hardly generate. The freedom of intensity modulation has a great impact on the dosimetric superiority of IMRT, in which better target coverage and less dose to the OARs can be achieved.

The superiority of IMRT over 3DCRT is illustrated in **Figure 2**, which shows radiotherapy plans for NPC patients. The dose-volume histogram (DVH) and the isodose distribution show that IMRT is more capable of sparing the dose delivered to both parotid glands while delivering an adequate dose to the PTV.

4. IMRT planning

To achieve the dosimetric superiority of IMRT described in the last section, the planning procedure adopts an inverse approach. Inverse planning is a process to determine the optimal beam intensity. Numerous inverse planning approaches have been proposed and they can be classified as dose-volume based or biological index based [30]. The inverse planning procedure starts with the delineation of the regions of interest (ROI) which includes the PTV and OAR, followed by the beam configuration, objective function setting, and computer optimization. The workflow of IMRT planning is illustrated in **Figure 3**.

The procedures which require human input, including the setting of ROI delineation, beam configuration, and objective function, and evaluation of the plan are further discussed in the following sections.

4.1 Target delineation

Target delineation is the first and a very important step in IMRT planning to ensure effective treatment. The delineation of targets in head and neck cancers includes the high-risk, intermediate-risk, and low-risk planning target volume (PTV) [31]. The

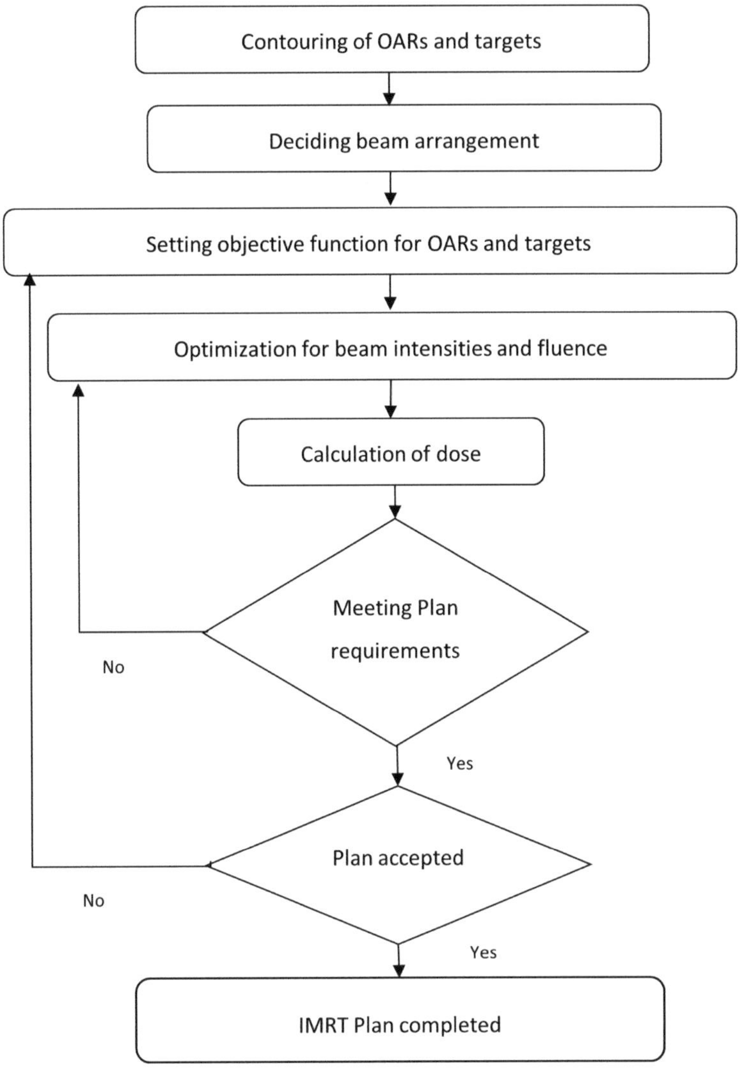

Figure 2.
Procedure of IMRT planning.

intermediate-risk PTV refers to the regional lymph nodes and the isotropic margins of the high-risk PTV, the low-risk PTV refers to selective negative lymph nodes for prophylactic treatment, and the high-risk PTV encompasses the primary tumor or tumor bed and the positive lymph nodes. The consensus guideline on the delineation of elective lymph nodes levels is well-established [32]. The guideline classifies the regional lymph nodes in the head and neck region into 10 levels and defines their anatomical boundaries. While the selection of lymph nodes levels to be treated largely depends on different oncologists' judgment and individual patients' conditions, there have been published guidelines to review the criteria for the lymph nodes levels selection for treatment in different types of head and neck cancers [32, 33]. Contrary to the well-established consensus in the delineation of PTV for the regional lymph

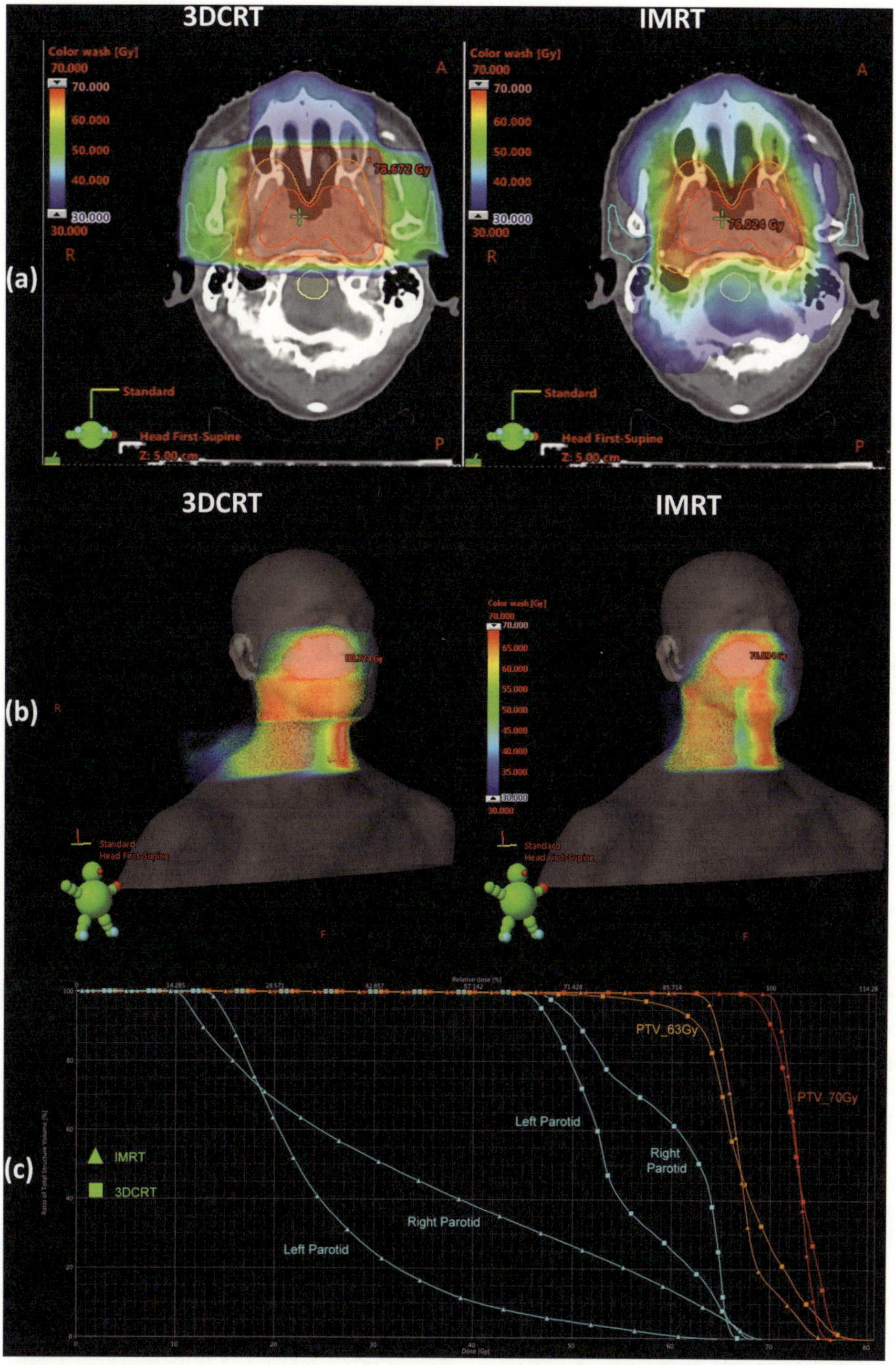

Figure 3.
Comparisons in NPC patients with 3DCRT and IMRT plans. (a) Isodose distribution; (b) 3-dimensional dose color wash; (c) dose-volume histogram.

nodes, the high-risk PTV delineation technique varies among oncologists. It can either be based on the isotropic expansion of the gross tumor volume or the inclusion of anatomical sub-sites [31]. The method of isotropic expansion to form PTV and the margins needed has been described [34]. The aim of the margins is to account for the uncertainties in the delivery of radiation to avoid target miss. On the other hand, the aim of the inclusion of anatomical subsites in the high-risk PTV in addition to the gross tumor volume is to include regions with possible microscopic extension [33].

The delineation of PTV is closely associated with the dose optimization regarding the skin dose. Usually, oncologists contour a clinical target volume (CTV) that covers all clinical and subclinical malignancy to be irradiated [35]. PTV, on the other hand, would add geometrical margins to CTV to ensure that the prescribed dose is adequately delivered. The CTV to PTV margins can be determined by previously reported margin recipes, accounting for systematic and random error during irradiation [36]. It is worth to note that there is a common circumstance when the head and neck cancers CTV stops just below the skin surface, i.e. no disease in the skin, while the PTV would cover the skin surface or even go beyond it after adding the CTV to PTV margins. In this case, the inverse planning procedure of IMRT would unnecessarily attempt to deliver an extra dose into the skin surface region [37], leading to excessive dose to the skin and adverse skin reactions [38]. Special attention is suggested to these cases, where the target is close to but not involving skin surface so PTV margins should be modified to avoid excessive skin surface normal tissue dose. Many imaging modalities contribute to the delineation of the target. It is important for the definition of tumor extent, the assessment of lymph nodes involvement, and the evaluation of perineural spread [39]. The common modalities include computed tomography (CT) and magnetic resonance imaging (MRI). Both CT and MRI are imaging modalities that provide sectional images with 3-dimensional reconstruction. Each of them has their unique strengths and therefore can provide complementary information in the localization of tumors and organs at risk.

Although both CT and MRI generate sectional images, their image generation mechanisms are not the same. The CT generates images using X-ray. By rotating the X-ray tube, a fan beam of X-ray is irradiated around the patients. After passing through the patient's body and being attenuated differentially by different body tissue with various densities, the X-ray detector receives many projections from the scanned body region. The computer then generates cross-sectional images based on the information gathered from the detected X-ray projections [40]. The resultant images are shown in grayscale according to the tissue density, which can be illustrated by appearing white for bone (high density), gray for soft tissue (medium density), and black for air (low density) [40]. In addition to the visualization of internal anatomy for the diagnosis purpose, the grayscale which is derived from the CT numbers and the robust geometrical information make the CT images suitable to be used for the dose calculation in radiotherapy planning [41].

On the other hand, MRI works by detecting the reaction of the MR-active nuclei in different parts of the body, mainly hydrogen, to the magnetic fields generated by the MRI machine [42]. MR-active nuclei refer to the particles that have net spins of the protons and neutrons, which create magnetic fields on the nuclei [43]. These MR-active nuclei, therefore, react to the strong magnetic field applied by the MRI machine. The image formation is first done by the application of magnetic field to patients' body to align the spinning axis of the MR-active nuclei in the body tissue. Then, by the application of short pulse radiofrequency, the alignment is displaced and then relaxed. This procedure, called relaxation, leads to the release of energy detected by the receiver coil [42, 44]. The two main types of relaxation are longitudinal relaxation time (T1) and transverse relaxation time (T2). T1 determines the rate of the spinning axis of the MR-active nuclei to realign to the MRI machine magnetic field, while T2 determines the rate of the MR-active nuclei to lose phase from the alignment [43]. The detection of the energy released can then be processed by computers to generate the cross-sectional images. The differences in the relaxation time (T1 or T2) and the density of the nuclei contribute to the tissue contrast in MRI images [43].

Utilization of both CT and MRI images in head and neck cancers is common because they are complementary to each other. In general, MRI is better in soft-tissue contrast while CT is better in detecting bone erosion. For example, T1 weighted MRI images are the most suitable to delineate NPC tumors because of better soft-tissue contrast and more sensitive in detecting the perineural extension of the tumor [45]. However, MRI images may fail to detect subtle skull base bone erosion, which can be complemented by coronary CT images in the bone window [46]. Also, in the cancer of the oral cavity, contrast-enhanced T1 weighted MRI images are the best for the delineation of tumor margin [47], while CT images are useful for the detection of the small lytic lesion in the cortical mandible [48].

In addition, PETCT also provides useful information to the commonly used CT and MRI images. The PETCT utilizes the mechanism of the increased uptake of the fluorodeoxyglucose (FDG) in tumor cells than in normal cells because of their higher metabolic activity [49]. The FDG uptake site can then be localized by scanners by detecting the radioactivity of the FDG. There are several circumstances that PETCT can provide supplementary information in addition to CT and MRI images. PETCT has been reported to have superior performance than CT and MRI in the detection of involved cervical lymph nodes. This is illustrated by the sensitivity of 90% and specificity of 94% in PETCT, compared with about 80% sensitivity and specificity in MRI and CT [50]. Also, PETCT is better in the detection of the unknown primary tumor, which is essential to decide the treatment regimen [51]. Furthermore, PETCT is useful in determining the presence of distant metastasis. It has the sensitivity and specificity of 89% and 95% respectively which indicates a very accurate diagnosis of the metastatic stage of the disease [52].

4.2 Organs at risk delineation

Inverse planning of IMRT involves the estimation of OAR dose for the calculation of the beam modulated intensity. The accuracy of the OARs delineation is crucial for the estimation of OARs dose, and hence the inverse planning procedure. There has been a consensus guideline on the OARs delineation in the head and neck regions [53]. This guideline listed the anatomical boundaries of 25 OARs in the head and neck region for the purpose of consistency in the delineation. Detailed atlas has also been supplemented for reference. **Figure 4** shows part of the atlas provided by the guideline

4.3 Beam arrangement

In the early application of IMRT, an equally spaced beam arrangement was commonly used [54, 55]. There are two other beam arrangement options available in the Eclipse treatment planning system (Varian Medical System, Palo Alto, USA). These include volumetric modulated arc therapy (VMAT) that enables rotational beams and beam angle optimization (BAO) that automatically chooses optimal static beam angles in either coplanar or non-coplanar beam arrangements.

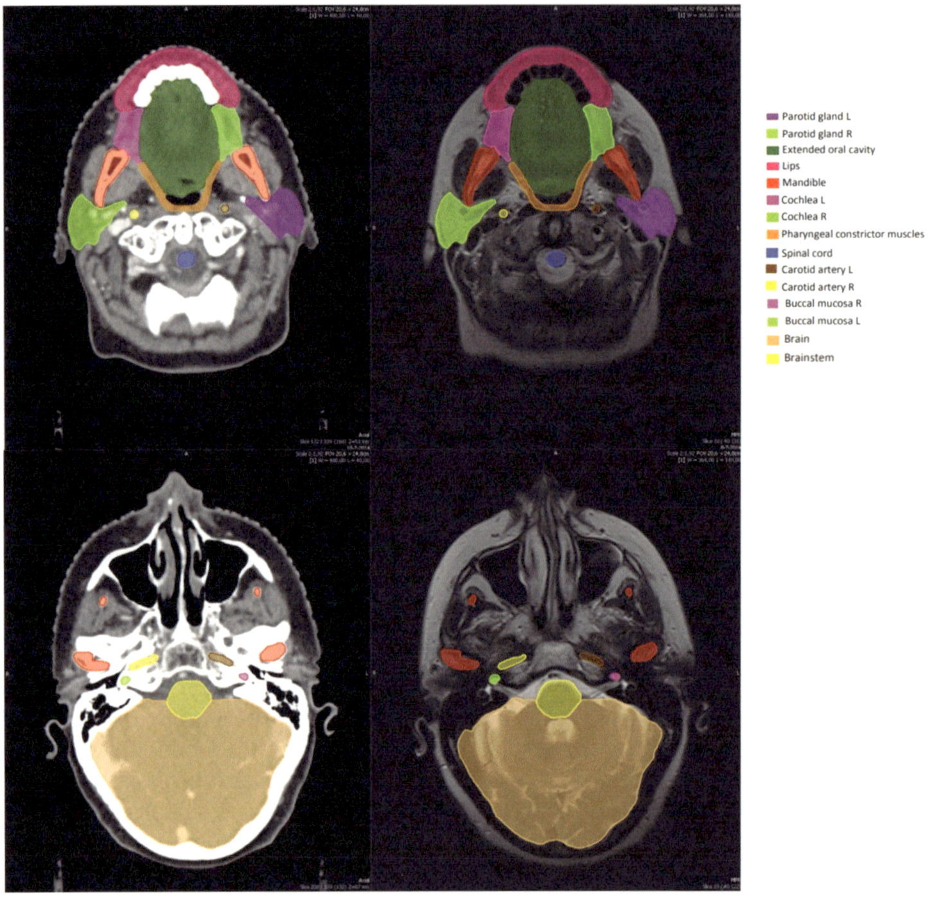

Figure 4.
Part of the OAR delineation atlas. Adapted from [53]. Copyright 2015 the Authors.

4.3.1 Equally spaced beams (ESB)

The delivery of IMRT requires several beams to achieve the assigned dose distribution [29]. It has been a common practice to use the 5–9 beams arrangement in IMRT for head and neck cancer [55, 56]. Theoretically, a greater number of beams can have a higher chance to achieve the planned dose distribution, which increases the time for delivery and quality assurance. Hence, effort should be put to minimize the number of beams to use. Another concern in the beam placement is that opposing beams should be avoided in IMRT because it reduces the effectiveness of the optimization [57]. Furthermore, it has been calculated that the optimal number of beams is 7–9 after striking a balance between the gain in dose distribution and the expenses of treatment time in further addition of beams [58].

4.3.2 Beam angle optimization (BAO)

Selecting optimal beam orientations can help to improve the dose distribution in complex plans [59]. BAO is a function available in the Eclipse treatment planning system that a built-in algorithm can automatically choose the optimal beam arrangements in static beam IMRT. The mechanism of selecting the beams is by elimination of beams from up to 400 pre-assigned beams orientations. Then, the calculation of fluence optimization iterations can help to eliminate the beams that cause the least contribution to the pre-set objective functions until the number of desired beams is reached. Planners must customize the resulting number of beams, coplanar or non-coplanar arrangement, and the number of initial beams. Also, objective functions for each target volume and OARs must be set beforehand for the purpose of fluence optimization in the beam elimination process. The user interface of BAO is shown in **Figure 5**.

4.3.3 Volumetric modulated arc therapy (VMAT)

VMAT is a technique that enables the delivery of IMRT in one or more rotations of the linear accelerator gantry. The delivery time is shorter than static gantry methods while maintaining at least comparable dosimetric quality [60]. It is done by simultaneous modulation of the position of the multi-leaf collimator (MLC), dose rate, and gantry speed, while the gantry is rotating around the patient during treatment. The VMAT plan optimization is done on the same user interface as the fixed beam IMRT plan, which is the photon optimizer in the Eclipse treatment planning system. While individual optimal fluence for the beam intensity modulation is optimized for the fixed beam IMRT, the VMAT optimization considers the full rotation of the gantry by dividing it into 178 equally spaced control points [61]. Assuming that the radiation

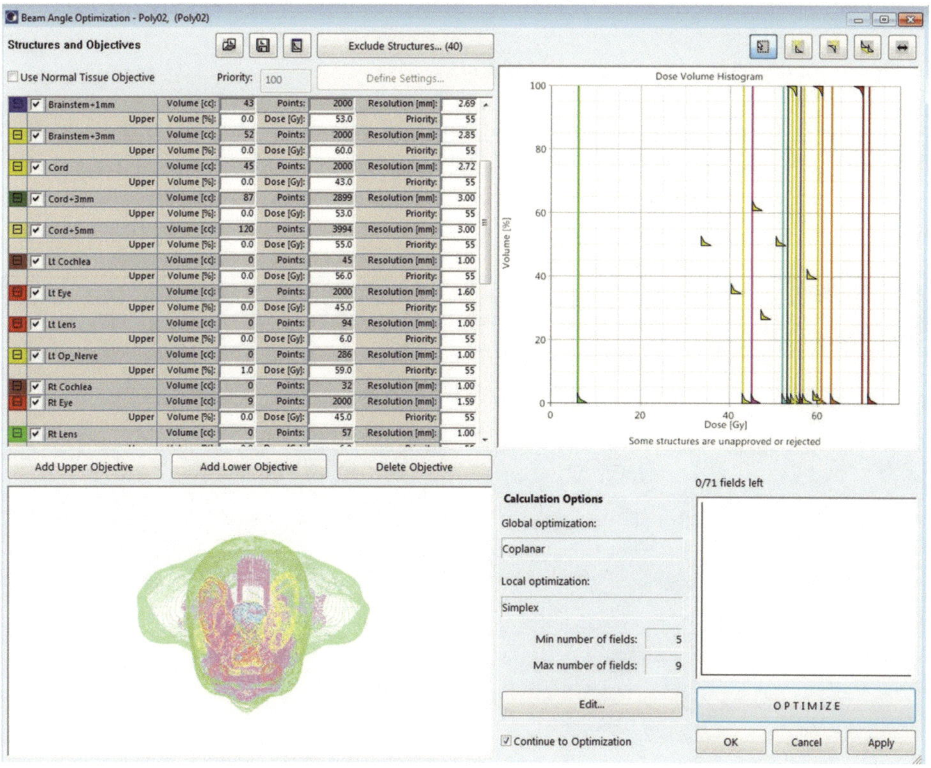

Figure 5.
User interface of BAO in Eclipse treatment planning system.

from each control point is delivered from a static gantry, the optimizer then generates the information of the MLC position, dose rate, and gantry speed altogether for the dose distribution calculation. The photon optimizer user interface for the optimization of IMRT in the Eclipse treatment planning system is shown in **Figure 6**.

4.4 Optimization objectives and procedures

The setting of dose objective is a crucial step in inverse planning because it defines the doses to be delivered to various delineated structures. The computer then calculates the intensity modulation of the treatment field based on the definition of dose objectives [62]. While both dose-volume based objectives and biological objectives can be input in the current commercially available system, dose-volume based objectives were more commonly used. This is because it has been demonstrated that the use of generalized equivalent uniform dose (gEUD) objectives would lead to poorer homogeneities [63]. Inverse planning was first proposed in 1982 [64], in which the dose distribution was defined by planners for the calculation of beam intensity to deliver the desired dose. It is an "inverse" process when compared with the conventional "forward" approach, in which the planners define beam parameters for the calculation of dose distribution [62]. There are upper objective, lower objective and mean objective in the definition of dose-volume based objectives for a structure. A priority number is assigned for each objective to indicate their relative importance. Because the objectives to achieve target dose coverage and to avoid dose to OARs sometimes oppose to each other, the setting of priority provides information for the computer system to decide the "trade-off" between conflicting objectives.

4.5 Dose constraints of targets and OARs

In general, there are 3 types of dose constraints settings before the optimization. They are the PTVs, serial OARS, and parallel OARs respectively. For the PTV, it requires the setting of at least one upper objective and one lower objective as shown in

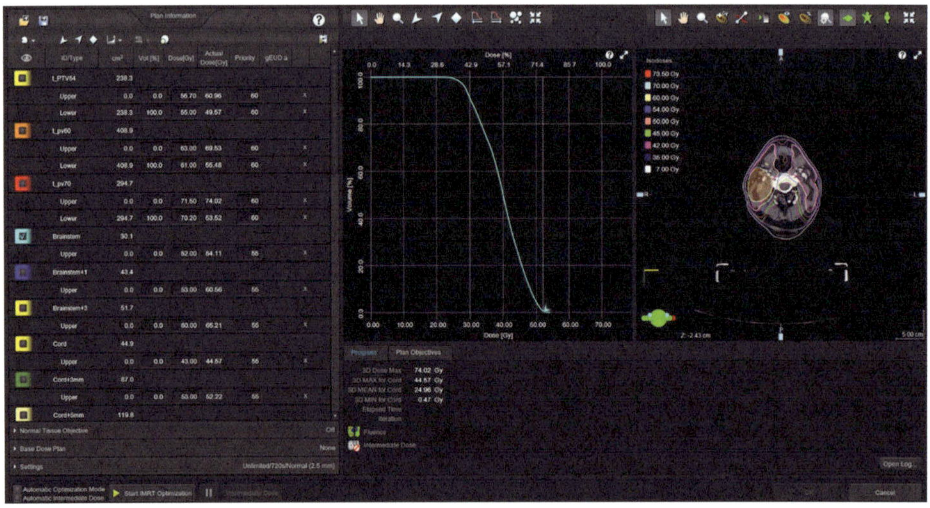

Figure 6.
User interface of photon optimizer.

Figure 7. The resultant dose-volume histogram (DVH) should show that the majority of the PTV receives the desired dose with little volume receive the higher dose, and the shape should look like a plateau at 100% volume with an extremely steep cliff at the end when it reaches the prescribed dose.

The dose constraints setting for serial OARs only requires an upper objective to limit its maximum dose, as shown in **Figure 8**.

For parallel OARs, since the dose received by the various proportion of volume is the concern for late side effects, setting of upper objectives to limit the maximum dose is not enough. It can be done by setting multiple upper objectives at different dose-volume levels or setting the mean objectives. The purpose is to limit the received dose at all volume levels and to push the DVH to its left end as much as possible. A sample objective setting for a parallel OAR is shown in **Figure 9**.

4.6 Practical difficulties of optimizing a radiotherapy plan for head and neck cancers

Although the planning procedures are driven by treatment planning computer calculations in an inverse planning process, it is not a completely automatic procedure and there are difficulties in the planning. The difficulties in planning are largely related to the number of OARs and the geometric relationship between the PTVs and the OARs. In the optimization process of the inverse planning, it is usually not possible to achieve all the lower objectives for the PTVs while fulfilling all the upper and mean objectives for the OARs because they naturally contradict each other when the PTVs and OARs are in the vicinity [65]. In head and neck cancers, there are many OARs near to the PTVs including but not limited to the brain stem, the spinal cord, the parotid gland, and the optic nerves. Because of this, the treatment planning system optimization usually has no optimal solution that can fulfill all the set objective functions. Therefore, planners need to intervene in the procedure by evaluating the optimized treatment plans using their own experiences, and to balance the trade-off among all the nonoptimal objective functions of the PTVs and OARs.

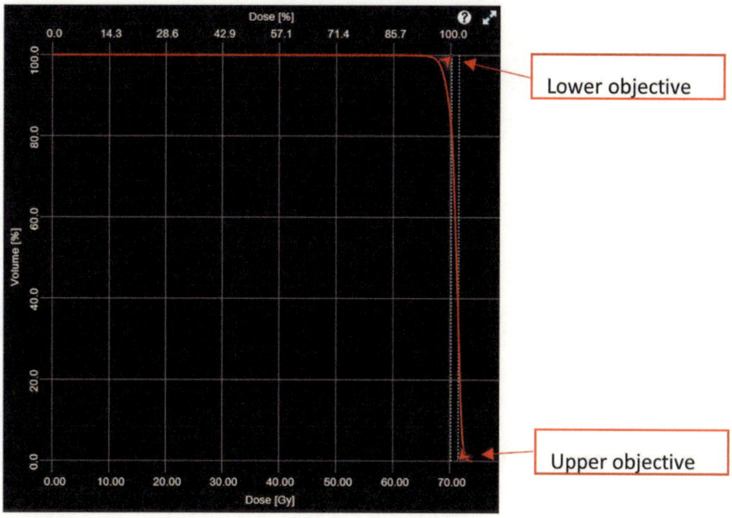

Figure 7.
Dose constraints setting of PTV.

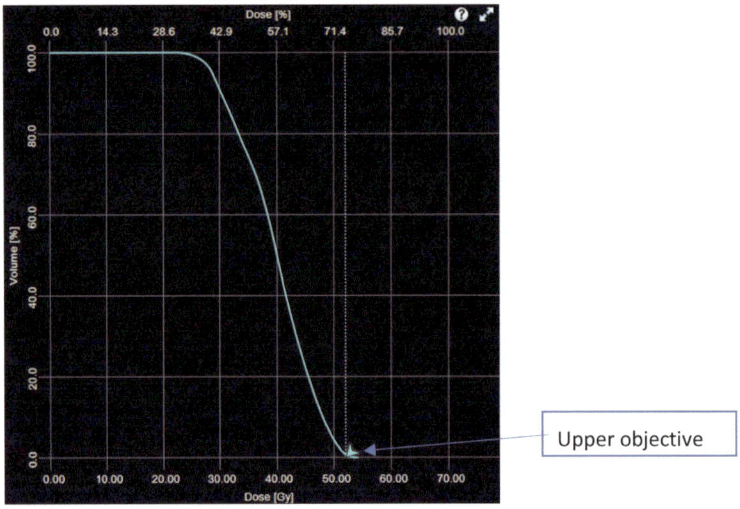

Figure 8.
Dose constraints setting of serial OARs.

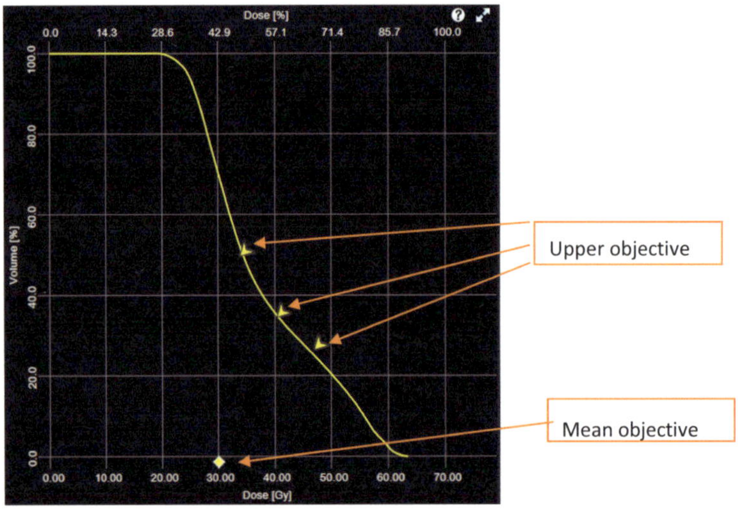

Figure 9.
Dose constraints setting of parallel OARs.

4.7 Plan evaluation

In the evaluation of radiotherapy plan dosimetric quality, there are four main parameters to be evaluated: (1) PTV coverage, (2) OAR dose, (3) PTV homogeneity, and (4) PTV conformity [66]. PTV coverage refers to the minimum proportion of PTV covered by the prescribed dose. OAR dose is to see whether it is within the organ tolerance. PTV homogeneity is used to assess the dose uniformity within the PTV whereas PTV conformity is to evaluate whether the prescribed dose level encompasses and follows the shape of the PTV. Examples of different PTV coverage, homogeneity, and conformity situations are illustrated in **Figure 10**.

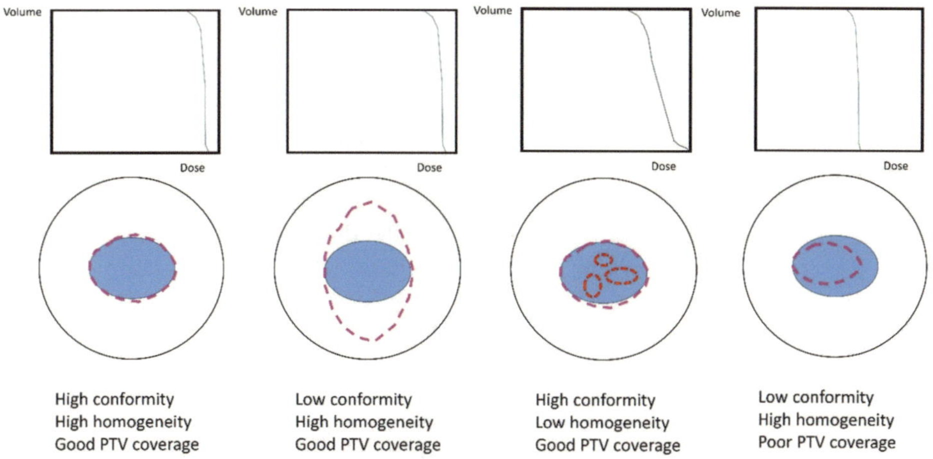

Figure 10.
Examples of different PTV coverage, homogeneity, and conformity situations. The PTV is in blue solid lines and the body is in black solid lines. The purple dashed lines are the prescribed isodose and the red dashed lines are the hot spots isodose. Their respective dose-volume histograms are shown above.

The evaluation of PTV coverage and OAR dose is conducted using the dose-volume histogram (DVH). PTV homogeneity and conformity are assessed by indices known as the homogeneity index [67] and conformity index respectively [68].

5. Current challenges and promises in head and neck cancer radiotherapy

As illustrated, IMRT offers the opportunity for better treatment outcome and less side effects in radiotherapy of head and neck cancers when compared with 3DCRT. A positive aspect of IMRT is that it can increase the dose conformity and homogeneity to the PTV while better sparing of the OARs [69, 70]. The following challenges are needed to be addressed for further development of the advantages of IMRT.

5.1 Organs at risk (OARs) dose estimation

In the treatment planning of IMRT, the inverse planning process requires planners to define the dose limits of various PTVs and OARs for the optimization of the beam intensity modulation. This process is regarded as the setting of the objective function, which includes the dose constraints and priority of the PTVs and OARs as discussed in Section 4.5. In general, the setting of PTVs objective functions are guided by the prescription whereas those for the OARs are set according to their dose tolerance [71]. In practice, however, the objectives for OARs sparing are often in conflict with the objectives to achieve PTV dose coverage [72]. This is because OARs and PTVs are often in close proximity and sometimes may even overlap one another. In this condition, we may have to deliver OARs doses that are close to or even higher than their dose tolerance in order to achieve PTV adequate dose coverage. On the contrary, when the OARs are far from the PTV, the actual OARs dose would be well below their tolerance. It is logical to deduce that the OARs dose is related to their anatomical relationship with PTVs, and this relationship varies greatly among different patients.

5.1.1 Knowledge-based radiotherapy and 4pi VMAT

Knowledge-based radiotherapy planning has recently emerged as rapidly developing area with the aim to improve the IMRT planning process [73]. Knowledge-based planning refers to the strategy to incorporate past plans data (known as knowledge) into the treatment planning process. Six different categories of purpose in knowledge-based planning have been summarized in a review article, which includes (1) the determination of DVH, (2) specific dose metrics, (3) voxel-level doses, (4) objective function weights, (5) beam parameters and (6) quality assurance metrics [73]. The development of knowledge-based radiotherapy planning enables planners to determine the setting of objective functions in a more systematic approach, less dependent on personal experience, and therefore higher consistency of plan qualities.

The technology of delivering 4pi VMAT is emerging. 4pi radiotherapy refers to the incorporation of beams distributed on the imaginary isotropically expanded spherical surface around the iso-center during plan optimization [74]. The 4pi VMAT can be delivered by non-coplanar arc beams using a static couch or synchronizing the arc rotation of the gantry with a rotating couch [75, 76]. It has been shown that 4pi VMAT has the potential to further decrease the dose to OARs compared with coplanar VMAT. For example, a study on head and neck cancers reported that the mean D_{max} to the brain stem and spinal were decreased by 6 Gy and 3.8 Gy respectively using 4pi VMAT [77]. In addition, the method of delivering 4pi VMAT with synchronized gantry and couch rotation enabled more sophisticated arc trajectories compared with the static couch method. It was expected to deliver a highly conformed dose to the PTV with a reduction of OARs dose and 50% isodose volume in the patient body [76]. Although the treatment time will increase by 30% in current linear accelerators compared with coplanar VMAT [75], the potential of 4pi VMAT can be unleashed with the advancement of the future linear accelerators with automatic couch and gantry motion capabilities for faster 4pi VMAT delivery [78].

5.2 Tumor dose escalation

IMRT offers the possibility to escalate the dose to the tumor because of its better ability to spare the OARs. In fact, dose-escalation has already been implemented in IMRT in the treatment of NPC when the gross tumor dose was raised from 66 Gy in conventional radiotherapy to about 70 Gy [79]. NPC is known for its radiosensitivity and the existence of dose-tumor-control relationship beyond routine cancericidal dose [80], hence increasing the dose to the tumor volume is able to increase the local control rate. It has been reported that in the group of predominantly locally advanced NPC (T3-4 N0-1), 61.8% of the failure was caused by local relapse [81]. Another study also revealed that 80% of the recurrent cases had the relapse sites at the region delivered with the median dose of 70.4 Gy in the previous treatment [82]. Clinical investigations on the dose escalation in the treatment of NPC using external beam radiotherapy [83] and brachytherapy have been reported [84]. Although it has shown good local control and survival in both reports, treatment side effects were the concern. For example, grade 3 mucositis was observed in about 80% of the cases [83]. Also, by assessing the acute toxicity, it has been suggested that the maximal tolerable dose in IMRT of head and neck cancers was 2.36 Gy per fraction to a total of 70.8 Gy [85].

5.2.1 Application of radiomics to selection of NPC cases for dose escalation

Radiomics refers to the extraction of features in the regions of interest (ROI) from medical images [86]. The extracted features can be the image voxel intensity, ROI texture and shape features, etc. [87]. These extracted radiomics features can be used to correlate with clinical data such as recurrence and metastasis status of patients, so as to develop tools for predicting treatment outcome in future patients based on individual patients' image radiomics features. Research articles have been published to evaluate the chance of local recurrence in NPC patients, and it was reported that local recurrence can be predicted using pre-treatment imaging with a concordance index of over 0.8 [88, 89]. The future direction could be to incorporate radiomics study for more accurate and individualized patient selection instead of based on their staging. With the attempt to generate own local recurrence prediction model based on radiomics features, NPC patients indicated for GTV dose escalation could be more accurately identified.

6. Summary

Radiotherapy is necessary for the treatment of various head and neck cancers either as a primary treatment or adjuvant treatment after surgery to cure the disease. To achieve optimal radiotherapy treatment, we need to understand the rationale of IMRT and the procedure of treatment planning. With the help of treatment planning computer, inverse planning procedure can accomplish treatment plans with highly conformal radiation dose to PTV and dose avoidance from OARs. Because of the conflicting nature of the 2 major dosimetric goals: high PTV dose and low nearby OARs dose, the optimal radiotherapy treatment is usually achieved by experienced planners who are able to carefully balance the trade-off between the conflicting goals. Nevertheless, the present development of knowledge-based planning could provide a guidance for planners to decide the trade-off in a more objective manner. In addition, the development of 4-pi VMAT and research of radiomics may strengthen the advantage of IMRT in terms of OARs sparing and tumor dose escalation.

Author details

Wan Shun Leung[1*] and Hing Ming Hung[2]

1 Department of Health Technology and Informatics, The Hong Kong Polytechnic University, Hong Kong

2 Department of Clinical Oncology, Pamela Youde Nethersole Eastern Hospital, Hong Kong

*Address all correspondence to: wsvleung@polyu.edu.hk

References

[1] Hong Kong Cancer Registry. Nasopharyngeal Cancer in 2016. 2016. Available from: http://www3.ha.org.hk/cancereg/facts.html

[2] Ng WT, Wong ECY, Lee VHF, Chan JYW, Lee AWM. Head and neck cancer in Hong Kong. Japanese Journal of Clinical Oncology. 2018;**48**(1):13-21

[3] Shield KD, Ferlay J, Jemal A, et al. The global incidence of lip, oral cavity, and pharyngeal cancers by subsite in 2012. CA: A Cancer Journal for Clinicians. 2017;**67**(1):51-64

[4] Ferlay J, Ervik M, Lam F. Nasopharynx fact sheet: GLOBOCAN 2018. 2018. Available from: http://gco.iarc.fr/today/data/factsheets/cancers/4-Nasopharynx-fact-sheet.pdf

[5] Chua MLKD, Wee JTSF, Hui EPF, Chan ATCP. Nasopharyngeal carcinoma. Lancet. 2015;**387**(10022):1012-1024

[6] Lee AWM, Ngan RKC, Tung SY, et al. Preliminary results of trial NPC-0501 evaluating the therapeutic gain by changing from concurrent-adjuvant to induction-concurrent chemoradiotherapy, changing from fluorouracil to capecitabine, and changing from conventional to accelerated radiotherapy fractionation in patients with locoregionally advanced nasopharyngeal carcinoma. Cancer. 2015;**121**(8):1328-1338

[7] Kam MK, Leung SF, Zee B, et al. Impact of intensity-modulated radiotherapy (IMRT) on salivary gland function in early-stage nasopharyngeal carcinoma (NPC) patients: A prospective randomized study. Journal of Clinical Oncology. 2005;**23**(Suppl. 16): 5501-5501

[8] Sze HCK, Ng AWY, Yuen KT, Lai JWY, Ng WT. Chapter 11—International consensus on delineation of target volumes and organs at risk. In: AWM L, Lung ML, Ng WT, editors. Nasopharyngeal Carcinoma. London, United Kingdom: Academic Press; 2019. pp. 239-261

[9] Chan ATC, Felip E. Nasopharyngeal cancer: ESMO clinical recommendations for diagnosis, treatment and follow-up. Annals of Oncology. 2009;**20**(Suppl. 4): iv123-iv125

[10] Barrett A, Dobbs J. In: Barrett A et al., editors. Practical Radiotherapy Planning. 4th ed. London: Hodder Arnold; 2009

[11] Mazeron J-J, Ardiet J-M, Haie-Méder C, et al. GEC-ESTRO recommendations for brachytherapy for head and neck squamous cell carcinomas. Radiotherapy and Oncology. 2009;**91**(2):150-156

[12] Ellis MA, Graboyes EM, Wahlquist AE, et al. Primary surgery vs radiotherapy for early stage oral cavity cancer. Otolaryngology and Head and Neck Surgery. 2018;**158**(4):649-659

[13] Fridman E, Na'ara S, Agarwal J, et al. The role of adjuvant treatment in early-stage oral cavity squamous cell carcinoma: An international collaborative study. Cancer. 2018;**124**(14):2948-2955

[14] Budach W, Bölke E, Kammers K, et al. Induction chemotherapy followed by concurrent radio-chemotherapy versus concurrent radio-chemotherapy alone as treatment of locally advanced squamous cell carcinoma of the head and neck (HNSCC): A meta-analysis of randomized trials. Radiotherapy and Oncology. 2015;**118**(2):238-243

[15] Huang S-H, O'Sullivan B. Oral cancer: Current role of radiotherapy and chemotherapy. Medicina Oral, Patología Oral y Cirugía Bucal. 2013;**18**(2):e233-e240

[16] Gomez DRMD, Zhung JEBA, Gomez JBA, et al. Intensity-modulated radiotherapy in postoperative treatment of oral cavity cancers. International Journal of Radiation Oncology, Biology, Physics. 2009;**73**(4):1096-1103

[17] Pfister DG, Laurie SA, Lefebvre J-L, et al. American society of clinical oncology clinical practice guideline for the use of larynx-preservation strategies in the treatment of laryngeal cancer. Journal of Clinical Oncology. 2006;**24**(22):3693-3704

[18] Bhalavat RL, Fakih AR, Mistry RC, Mahantshetty U. Radical radiation vs surgery plus post-operative radiation in advanced (resectable) supraglottic larynx and pyriform sinus cancers: A prospective randomized study. European Journal of Surgical Oncology. 2003;**29**(9):750-756

[19] Timme DW, Jonnalagadda S, Patel R, Rao K, Robbins KT. Treatment selection for T3/T4a laryngeal cancer: Chemoradiation versus primary surgery. The Annals of Otology, Rhinology, and Laryngology. 2015;**124**(11):845-851

[20] Bristol IJMD, Ahamad AMD, Garden ASMD, et al. Postoperative radiotherapy for maxillary sinus cancer: Long-term outcomes and toxicities of treatment. International Journal of Radiation Oncology, Biology, Physics. 2007;**68**(3):719-730

[21] Skóra T, Nowak-Sadzikowska J, Mucha-Małecka A, Szyszka-Charewicz B, Jakubowicz J, Gliński B. Postoperative irradiation in patients with pT3-4N0 laryngeal cancer: Results and prognostic factors. European Archives of Oto-Rhino-Laryngology. 2015;**272**(3):673-679

[22] Wolf GT, Fisher SG, Hong WK, et al. Induction chemotherapy plus radiation compared with surgery plus radiation in patients with advanced laryngeal cancer. The New England Journal of Medicine. 1991;**324**(24):1685-1690

[23] Won HS, Chun SH, Kim B-S, et al. Treatment outcome of maxillary sinus cancer. Rare Tumors. 2009;**1**(2): e36-e114

[24] Katz TS, Mendenhall WM, Morris CG, Amdur RJ, Hinerman RW, Villaret DB. Malignant tumors of the nasal cavity and paranasal sinuses. Head & Neck. 2002;**24**(9):821-829

[25] Huang D, Xia P, Akazawa P, et al. Comparison of treatment plans using intensity-modulated radiotherapy and three-dimensional conformal radiotherapy for paranasal sinus carcinoma. International Journal of Radiation Oncology, Biology, Physics. 2003;**56**(1):158-168

[26] Adelstein DJMD, Koyfman SAMD, El-Naggar AKMDP, Hanna EYMD. Biology and management of salivary gland cancers. Seminars in Radiation Oncology. 2012;**22**(3):245-253

[27] Spratt DE, Salgado LR, Riaz N, et al. Results of photon radiotherapy for unresectable salivary gland tumors: Is neutron radiotherapy's local control superior? Radiology and Oncology. 2014;**48**(1):56-61

[28] Schoenfeld JDMD, Sher DJMDMPH, Norris CMMD, et al. Salivary gland tumors treated with adjuvant intensity-modulated radiotherapy with or without concurrent chemotherapy. International Journal of Radiation Oncology, Biology, Physics. 2012;**82**(1):308-314

[29] Brahme A. Optimization of stationary and moving beam radiation therapy techniques. Radiotherapy and Oncology. 1988;**12**(2):129-140

[30] Chui C-S, Spirou SV. Inverse planning algorithms for external beam radiation therapy. Medical Dosimetry. 2001;**26**(2):189-197

[31] Elicin O, Terribilini D, Shelan M, et al. Primary tumor volume delineation in head and neck cancer: Missing the tip of the iceberg? Radiation Oncology. 2017;**12**(1):102-102

[32] Grégoire V, Ang K, Budach W, et al. Delineation of the neck node levels for head and neck tumors: A 2013 update. DAHANCA, EORTC, HKNPCSG, NCIC CTG, NCRI, RTOG, TROG consensus guidelines. Radiotherapy and Oncology. 2014;**110**(1):172-181

[33] Eisbruch A, Foote RL, O'Sullivan B, Beitler JJ, Vikram B. Intensity-modulated radiation therapy for head and neck cancer: Emphasis on the selection and delineation of the targets. Seminars in Radiation Oncology. 2002;**12**(3):238-249

[34] Antolak JA, Rosen II. Planning target volumes for radiotherapy: How much margin is needed? International Journal of Radiation Oncology, Biology, Physics. 1999;**44**(5):1165-1170

[35] International Commission on Radiation Units and Measurements. Prescribing, Recording, and Reporting Photon Beam Therapy. Bethesda, MD: International Commission on Radiation Units and Measurements; 1993

[36] McKenzie AL, Herk MV, Mijnheer B. The width of margins in radiotherapy treatment plans. Physics in Medicine and Biology. 2000;**45**(11):3331-3342

[37] Thomas SJ, Hoole ACF. The effect of optimization on surface dose in intensity modulated radiotherapy (IMRT). Physics in Medicine and Biology. 2004;**49**(21):4919-4928

[38] Lee N, Chuang C, Quivey JM, et al. Skin toxicity due to intensity-modulated radiotherapy for head-and-neck carcinoma. International Journal of Radiation Oncology, Biology, Physics. 2002;**53**(3):630-637

[39] Rumboldt Z, Gordon L, Gordon L, Bonsall R, Ackermann S. Imaging in head and neck cancer. Current Treatment Options in Oncology. 2006;**7**(1):23-34

[40] Seeram E. Computed Tomography: Physical Principles, Clinical Applications, and Quality Control. 4th ed. St. Louis, Missouri: Elsevier; 2016

[41] Parker RP, Hobday PA, Cassell KJ, Sank VJ. The direct use of ct numbers in radiotherapy dosage calculations for inhomogeneous media. Journal of Computer Assisted Tomography. 1980;**4**(1):136

[42] Grover VPB, Tognarelli JM, Crossey MME, Cox IJ, Taylor-Robinson SD, McPhail MJW. Magnetic resonance imaging: Principles and techniques: Lessons for clinicians. Journal of Clinical and Experimental Hepatology. 2015;**5**(3):246-255

[43] Bitar R, Leung G, Perng R, et al. MR pulse sequences: What every radiologist wants to know but is afraid to ask. Radiographics. 2006;**26**(2):513-537

[44] Westbrook C, Talbot J. MRI in Practice. 5th ed. Newark: Wiley; 2018

[45] Rumboldt Z, Castillo M, Smith JK. The palatovaginal canal: Can it be identified on routine CT and MR imaging? AJR American Journal of Roentgenology. 2002;**179**(1):267-272

[46] K-i S, Hareyama M, Tamakawa M, et al. Prognostic factors of nasopharynx tumors investigated by MR imaging and the value of MR imaging in the newly published TNM staging. International Journal of Radiation Oncology, Biology, Physics. 1999;**43**(2):273-278

[47] Lam P, Au-Yeung KM, Cheng PW, et al. Correlating MRI and histologic tumor thickness in the assessment of oral tongue cancer. AJR American Journal of Roentgenology. 2004;**182**(3):803-808

[48] Mukherji SK, Isaacs DL, Creager A, Shockley W, Weissler M, Armao D. CT detection of mandibular invasion by squamous cell carcinoma of the oral cavity. AJR American Journal of Roentgenology. 2001;**177**(1):237-243

[49] Berger A. How does it work? Positron emission tomography. BMJ. 2003;**326**(7404):1449-1449

[50] Adams S, Baum RP, Stuckensen T, Bitter K, Hör G. Prospective comparison of 18F-FDG PET with conventional imaging modalities (CT, MRI, US) in lymph node staging of head and neck cancer. European Journal of Nuclear Medicine. 1998;**25**(9):1255-1260

[51] Kwee TC, Kwee RM. Combined FDG-PET/CT for the detection of unknown primary tumors: Systematic review and meta-analysis. European Radiology. 2008;**19**(3):731-744

[52] Xu G-Z, Guan D-J, He Z-Y. (18) FDG-PET/CT for detecting distant metastases and second primary cancers in patients with head and neck cancer. A meta-analysis. Oral Oncology. 2011;**47**(7):560-565

[53] Brouwer CL, Steenbakkers RJHM, Bourhis J, et al. CT-based delineation of organs at risk in the head and neck region: DAHANCA, EORTC, GORTEC, HKNPCSG, NCIC CTG, NCRI, NRG oncology and TROG consensus guidelines. Radiotherapy and Oncology. 2015;**117**(1):83-90

[54] Vlachaki MT, Teslow TN, Amosson C, Uy NW, Ahmad S. IMRT versus conventional 3DCRT on prostate and normal tissue dosimetry using an endorectal balloon for prostate immobilization. Medical Dosimetry. 2005;**30**(2):69-75

[55] Gupta T, Agarwal J, Jain S, et al. Three-dimensional conformal radiotherapy (3D-CRT) versus intensity modulated radiation therapy (IMRT) in squamous cell carcinoma of the head and neck: A randomized controlled trial. Radiotherapy and Oncology. 2012;**104**(3):343-348

[56] Ahmed M, Hansen VN, Harrington KJ, Nutting CM. Reducing the risk of xerostomia and mandibular osteoradionecrosis: The potential benefits of intensity modulated radiotherapy in advanced oral cavity carcinoma. Medical Dosimetry. 2009;**34**(3):217-224

[57] Soyfer V, Meir Y, Corn BW, et al. AP-PA field orientation followed by IMRT reduces lung exposure in comparison to conventional 3D conformal and sole IMRT in centrally located lung tumors. Radiation Oncology. 2012;**7**(1):23-23

[58] Webb S. Optimizing the planning of intensity-modulated radiotherapy. Physics in Medicine and Biology. 1994;**39**(12):2229-2246

[59] Stein J, Mohan R, Wang X-H, et al. Number and orientations of beams in intensity-modulated radiation treatments. Medical Physics. 1997;**24**(2):149-160

[60] Vanetti E, Clivio A, Nicolini G, et al. Volumetric modulated arc radiotherapy for carcinomas of the oro-pharynx, hypo-pharynx and larynx: A treatment planning comparison with fixed field IMRT. Radiotherapy and Oncology. 2009;**92**(1):111-117

[61] Vanetti E, Nicolini G, Nord J, et al. On the role of the optimization algorithm of RapidArc® volumetric modulated arc therapy on plan quality and efficiency. Medical Physics. 2011;**38**(11):5844-5856

[62] Cho B. Intensity-modulated radiation therapy: A review with a physics perspective. Radiation Oncology Journal. 2018;**36**(1):1-10

[63] Wu Q, Djajaputra D, Wu Y, Zhou J, Liu HH, Mohan R. Intensity-modulated radiotherapy optimization with gEUD-guided dose–volume objectives. Physics in Medicine and Biology. 2003;**48**(3):279-291

[64] Brahme A, Roos JE, Lax I. Solution of an integral equation encountered in rotation therapy. Physics in Medicine and Biology. 1982;**27**(10):1221-1229

[65] Tanaka Y, Fujimoto K, Yoshinaga T. Dose-volume constrained optimization in intensity-modulated radiation therapy treatment planning. Journal of Inequalities and Applications. 2015;**2015**(1):1-13

[66] Funk RK, Stockham AL, Laack NN. Basics of radiation therapy. In: Clinical Cardio-oncology. Amsterdam TN: Elsevier Inc.; 2016. pp. 39-60

[67] Grégoire V, Mackie TR. State of the art on dose prescription, reporting and recording in intensity-modulated radiation therapy (ICRU report No. 83). Cancer Radiothérapie. 2011;**15**(6):555-559

[68] Avt R, Mak ACA, Moerland MA, Elders LH, van der Zee W. A conformation number to quantify the degree of conformality in brachytherapy and external beam irradiation: Application to the prostate. International Journal of Radiation Oncology, Biology, Physics. 1997;**37**(3):731-736

[69] Zhang B, Mo Z, Du W, Wang Y, Liu L, Wei Y. Intensity-modulated radiation therapy versus 2D-RT or 3D-CRT for the treatment of nasopharyngeal carcinoma: A systematic review and meta-analysis. Oral Oncology. 2015;**51**(11):1041-1046

[70] Daly ME, Le Q-T, Jain AK, et al. Intensity-modulated radiotherapy for locally advanced cancers of the larynx and hypopharynx. Head & Neck. 2011;**33**(1):103-111

[71] Brodin NP, Tomé WA. Revisiting the dose constraints for head and neck OARs in the current era of IMRT. Oral Oncology. 2018;**86**:8-18

[72] Banaei A, Hashemi B, Bakhshandeh M, Mofid B. Trade-off between the conflicting planning goals in correlation with patient's anatomical parameters for intensity-modulated radiotherapy of prostate cancer patients. Journal of Radiotherapy in Practice. 2019;**18**(3):232-238

[73] Ge Y, Wu QJ. Knowledge-based planning for intensity-modulated radiation therapy: A review of data-driven approaches. Medical Physics. 2019;**46**(6):2760-2775

[74] Tran A, Zhang J, Woods K, et al. Treatment planning comparison of IMPT, VMAT and 4π radiotherapy for prostate cases. Radiation Oncology. 2017;**12**(1):10-10

[75] Wild E, Bangert M, Nill S, Oelfke U. Noncoplanar VMAT for nasopharyngeal

tumors: Plan quality versus treatment time: Noncoplanar VMAT for nasopharyngeal tumors. Medical Physics (Lancaster). 2015;**42**(5):2157-2168

[76] Lyu Q, Yu VY, Ruan D, Neph R, O'Connor D, Sheng K. A novel optimization framework for VMAT with dynamic gantry couch rotation. Physics in Medicine and Biology. 2018;**63**(12):125013-125013

[77] Subramanian VS, Subramani V, Chilukuri S, et al. Multi-isocentric 4π volumetric-modulated arc therapy approach for head and neck cancer. Journal of Applied Clinical Medical Physics. 2017;**18**(5):293-300

[78] Khan SJ, Chin E, Otto K, Hristov DH, Xing L, Fahimian BP. Beyond VMAT—assessing the potential of noncoplanar arc delivery trajectories incorporating dynamic couch motion in intracranial radiation therapy. International Journal of Radiation Oncology, Biology, Physics. 2016;**96**(2):S80-S81

[79] Kam MKM, Teo PML, Chau RMC, et al. Treatment of nasopharyngeal carcinoma with intensity-modulated radiotherapy: The Hong Kong experience. International Journal of Radiation Oncology, Biology, Physics. 2004;**60**(5):1440-1450

[80] Teo PML, Leung SF, Lee WY, Zee B. Intracavitary brachytherapy significantly enhances local control of early T-stage nasopharyngeal carcinoma: The existence of a dose–tumor-control relationship above conventional tumoricidal dose. International Journal of Radiation Oncology, Biology, Physics. 2000;**46**(2):445-458

[81] Chua DTT, Sham JST, Wei WI, Ho WK, Au GKH. The predictive value of the 1997 American Joint Committee on Cancer stage classification in determining failure patterns in nasopharyngeal carcinoma. Cancer. 2001;**92**(11):2845-2855

[82] Dawson LA, Anzai Y, Marsh L, et al. Patterns of local-regional recurrence following parotid-sparing conformal and segmental intensity-modulated radiotherapy for head and neck cancer. International Journal of Radiation Oncology, Biology, Physics. 2000;**46**(5):1117-1126

[83] Kwong DLW, Sham JST, Leung LHT, et al. Preliminary results of radiation dose escalation for locally advanced nasopharyngeal carcinoma. International Journal of Radiation Oncology, Biology, Physics. 2006;**64**(2):374-381

[84] Chao H-L, Liu S-C, Tsao C-C, et al. Dose escalation via brachytherapy boost for nasopharyngeal carcinoma in the era of intensity-modulated radiation therapy and combined chemotherapy. Journal of Radiation Research. 2017;**58**(5):654-660

[85] Lauve A, Morris M, Schmidt-Ullrich R, et al. Simultaneous integrated boost intensity-modulated radiotherapy for locally advanced head-and-neck squamous cell carcinomas: II—clinical results. International Journal of Radiation Oncology, Biology, Physics. 2004;**60**(2):374-387

[86] Larue RTHM, Defraene G, De Ruysscher D, Lambin P, Van Elmpt W. Quantitative radiomics studies for tissue characterization: A review of technology and methodological procedures. The British Journal of Radiology. 2017;**90**(1070):20160665-20160665

[87] Aerts HJWL, Velazquez ER, Leijenaar RTH, et al. Decoding tumour phenotype by noninvasive imaging using a quantitative radiomics approach. Nature Communications. 2014;**5**

[88] Zhang L, Zhou H, Gu D, et al. Radiomic nomogram: Pretreatment evaluation of local recurrence in nasopharyngeal carcinoma based on MR imaging. Journal of Cancer. 2019;**10**(18):4217-4225

[89] Zhang L-L, Huang M-Y, Li Y, et al. Pretreatment MRI radiomics analysis allows for reliable prediction of local recurrence in non-metastatic T4 nasopharyngeal carcinoma. eBioMedicine. 2019;**42**:270-280

Chapter 3

Parameters Affecting Pre-Treatment Dosimetry Verification

E. Ishmael Parsai and Elahheh Salari

Abstract

To assure the accuracy and safety of radiation delivery, it is highly recommended to perform pretreatment verification for complex treatment methods such as intensity-modulated radiation therapy (IMRT) or volumetric-modulated arc therapy (VMAT) to detect any potential errors in the treatment planning process and machine deliverability. It is expected that a qualified medical physicist is aware of the underlying scientific principles of imaging and therapeutic processes to perform or supervise technical aspects of pretreatment procedures to ensure safe and effective delivery of the treatment. For this purpose, several guidelines have been published to help direct medical physicists to evaluate the accuracy of treatment planning system (TPS) in the calculation of radiation dose, and dosimetry equipment to avoid possible errors. This will require a clear understanding of abilities as well as the limitations of each TPS, the dosimetry equipment at hand, and the gamma index to perform a comprehensive pre-treatment verification.

Keywords: pre-treatment verification, gamma index, treatment planning algorithms, beam modeling, detector resolution, planned dose grid, modulation index

1. Introduction

As a treatment modality driven by technology, radiation therapy (RT) has made significant advances in recent years. These advances have mostly been in areas of treatment delivery, imaging, and image fusion which has required sophisticated algorithms for calculation of dose in patients and complex machines to deliver the dose. There is always some level of discrepancies between the calculated dose and delivered dose which can arise from different sources such as: the dose calculation algorithm, beam modeling in TPS, physics data entry, beam delivery, detector resolution, and planned grid size. The purpose of this chapter is to present a review of algorithms for photon dose calculation, beam modeling in different TPSs, detector resolution and planned grid size (GS) and analyze the effect of each of them on gamma passing rate (GPR) in pre-treatment quality assurance (QA).

2. Treatment planning system (TPS) algorithms

For understanding TPS algorithms, it is required to know [1]:

1. The production of Megavoltage X-rays

2. The interaction and scattering of photons by the Compton effect

3. The effects of transport of charged particles near boundaries and tissue heterogeneities

By far, medical linear accelerators (linac) are the main devices used in the treatment of cancer patients producing X-rays and electrons in the clinical energy range. In the head of a linac, high energy electrons are accelerated to the near speed of light and are directed to strike a high Z target typically made of Tungsten which has also a high melting point to produce photon. The bremsstrahlung photons produced by a linac have an energy distribution from 0 to maximum energy of the electrons in the beam impinging upon the target. These photons pass through the primary collimator and other parts of the linac head such as jaws, Multi Leaf Collimator (MLC) system, etc. before reaching the patient. All these photons (primary and scattered) will contribute to photon fluence. For example, for a typical Varian linac with a flattening filter, 80–90% of primary photons are directly from the target, 3–5% from the primary collimator, 8–12% originated from the flattening filter [1, 2]. However, in modern linacs which are equipped with flattening filter free (FFF) technology, scatter photon produced in the treatment head has significantly decreased [3, 4]. Therefore, the contribution of primary and scatter photon in photon fluence for FFF is different from the flattened beam [5]. For example for a $40 \times 40\ cm^2$ field size and a 6 MV FFF beam, the calculated contribution was 84.6% for the primary source, 11.3% for the first scattered source, and 4.1% for the second scattered source [5].

In general, ionizing radiation such as photon, electron, and heavy charged particles interact with matter which depends on the energy of ionizing radiation, type of ionizing radiation, the atomic number, and density of the medium through which they travel. Photons are indirect ionizing radiation and energy deposition of the photon to the material is dominated by three interactions: Photoelectric, Compton scatter, and pair production. In the energy range from 100 Kev to 10 MeV, which is a mostly therapeutic range, the Compton process is dominant for energy absorption in soft tissues. The energy deposition of photons involves two stages: First, partial transfer of their kinetic energy to charged particles (electron, positron) when they interact with material, and second, energy deposition from these charged particles to material through excitation or ionization. The range of charged particles in the therapeutic energy range can be several centimeters so they can travel and pass-through various layers with different densities and atomic numbers in a human body. When charged particle equilibrium (CPE) is achieved, then there is a linear relationship between TERMA[1] (Total energy released per unit mass) and dose, and the two steps can be included in a single calculation. However, this condition does not occur near the edge of the field or in inhomogeneity regions like at tissue interfaces, therefore, this

[1] the production of mass attenuation coefficient and primary energy fluence of photon.

simplification cannot be valid, and the two steps of energy deposition of the photon to medium must be more clearly distinguished [6].

The human body consists of a variety of tissues and cavities that are radiologically different from water, such as lungs, oral cavities, teeth, nasal passages, sinuses, and bones. A treatment planning system uses the electron density derived from CT images of patients to calculate dose in the patient body. Therefore, the dose distribution inside the patient body is affected by these heterogeneities. In this area, the ability of treatment planning systems to calculate dose at the interferences such as lung vs. tissue, bone vs. air cavity, etc. is crucial. Also, using CT images with 3D TPS allows us to design a plan with complex beam arrangements which require more advanced dose computation algorithms. In this section, we will present a summary review of the past and current dose calculation algorithms used in the TPS for radiotherapy.

According to the American Association of Physicists in Medicine (AAPM) Task Group 65 (TG-65, Report No.85) [6], there are four types of inhomogeneity correction algorithms:

Category.1: Linear attenuation, Ratio of tissue air ratio (RTAR), Power law (Batho).

Category.2: Equivalent TAR, Differential scatter air ratio (dSAR), Delta volume, Differential TAR, and 3D Beam subtraction method.

Category.3: Convolution (pencil beam) and Fast Fourier transformation (FFT) techniques.

Category.4: Superposition/Convolution, Monte Carlo.

2.1 Category.1

2.1.1 Linear attenuation

This is the simplest technique for computation of inhomogeneity correction factor (ICF), which does not include any information regarding electron density and the geometric treatment beam parameters such as field size [6].

$$ICF = (\%\text{per } cm) \times inhomogeneity\ thickness\ (cm) \tag{1}$$

2.1.2 Ratio of tissue air ratio (RTAR)

Only heterogeneity correction applied on the beam path from source to the calculation point.

$$ICF(d, r) = \frac{TAR(d', r)}{TAR(d, r)} \tag{2}$$

where d and d' are physical depth and water equivalent depth to the calculation point and r is the field size at depth d. The main weakness of this method is overcorrection when the density of the medium is less than the density of the water and under correction when the density is greater than the density of water due to compromised modeling of lateral component of the scattered photon [7].

2.1.3 Power law (Batho)

This is an empirical correction factor method for points lying within water and distal to an inhomogeneity by raising tissue-air ratios to a power that depends on density. This was first proposed by Batho in 1964 [8] and then modified by Sontag and Cunningham in 1977 [9].

$$ICF = \frac{TAR(d_1, r)^{\rho_1 - \rho_2}}{TAR(d_2, r)^{1-\rho_2}} \tag{3}$$

where d_1 is depth to first slab boundary and d_2 is depth to second slab boundary from the point of calculation at depth d. r is field size at depth d and ρ_1 and ρ_2 are densities of the medium in which the calculation point is located and relative electron density of the overlying material respectively.

The power law method underestimates the dose when density is less than one and overestimates when density is greater than one [6]. Several studies showed improvement if Tissue Maximum Ratio (TMR) is used instead of TAR [10, 11].

2.2 Category.2

2.2.1 Equivalent TAR (ETAR)

It can be considered as the first practical dose calculation method using the full CT data set for computerized treatment planning and was used in early treatment planning systems [6].

$$ICF(d, r) = \frac{TAR(d', \tilde{r})}{TAR(d, r)} \tag{4}$$

where d' and $\tilde{r}$ represent the "scaled" or "effective" values of depth at interesting point (d) and field radius (r) respectively for the energy of the radiation being used. This method required excessive computer memory and calculation times; therefore, some adjustments such as the coalescing of adjacent CT slices were applied to reduce 3D calculations to appropriate 2D calculations to make it more practical for use in clinics in the 1980s.

2.2.2 Differential scatter air ratio (dSAR)

This was a 3D dose calculation in a heterogeneous media that used scatter-air ratios (SAR) to calculate the dose to a point in an inhomogeneous medium. For this purpose, a SAR table was used to determine the scatter contribution that arises from voxels within the irradiation volume [12].

2.2.3 Delta volume (DVOL)

The primary dose, an analytical first-scatter dose component, and an approximate residual multiple-scatter component were summed to calculate dose at a point in a heterogeneous medium. This method has been examined and justified for Co-60 and succeeds incorrectly calculating the dose to (a) water with a small void and, (b) homogeneous non-water medium.

dSAR and DVOL have never been implemented in clinics due to the long ***CPU*** time required to run them with no significant improvement in dose calculation accuracy compared to previously used algorithms [7].

2.2.4 Differential TAR

Kappas and Rosenwald [13] showed that applying K(θ,μ) on dSAR method results in more accurate results.

$$K(\theta, \mu) = e^{(\mu_0 cos\theta - \mu_1(\theta))(\overline{b} - b)} \tag{5}$$

where μ_0 and μ_1 are the linear attenuation coefficients in the water of the primary and of the first-order scattered photons arriving at a point after a deflection. $\overline{b}$ is the path length en route to point (in the waterlike medium) and b is the corresponding effective path length (in the heterogeneous medium). For very large fields and depths and when the thickness of the overlying tissue is greater than 5 cm, the difference between measurement and calculation is more than 2% and less than 6% [6].

In general, categories 1 and 2 are not applicable when photon energy is greater than 6MV where scatter contribution is less important, and the effects of secondary electrons (delta rays) set in motion can result in very high local dose changes [6].

2.3 Category.3

2.3.1 Convolution techniques

This technique is a model-based algorithm which unlike correction-based algorithms uses heterogeneity effects directly to compute the dose in tissue. Kernels are used for modeling the dose distribution in media. The kernels represent the energy spread and dose deposition of secondary particles from an interaction at a given point or line which is not usually accessible through measurements but is very simple to calculate by use of Monte Carlo particle transport codes [12]. Absorbed dose is calculated based on the following equation

$$Absorbed\ Dose = energy\ fluence\ distribution \otimes \mathrm{K} \tag{6}$$

This means that the energy fluence distribution is ***convolved***[2] with the scatter spread kernel (K) to obtain the dose.

Energy deposition Kernel (EDK) is the energy distribution revealed to volume elements (per unit volume) in an irradiated medium, commonly water. There are three different categories for EDKs based on the geometry of the elemental beam that delivers the incident energy: A point kernel, pencil kernel, and planar kernel [7].

Point Kernel: This kernel describes the pattern of energy deposition in an infinite media around a primary photon interaction site.

Pencil Kernel: This kernel describes the energy deposition in a semi-infinite medium from a point monodirectional beam.

Planar Kernel: A planar kernel describes the energy spread from primary interactions located in a plane of an infinite broad beam.

In 1986, Mohan et al. [14] introduced a ***differential Pencil beam algorithm*** which is a good example of this category. This is the simplest and fastest algorithm for dose calculation because it only considers inhomogeneity corrections in longitudinal

---

[2] Convolution, ⊗, is a mathematical operation used to combine functions.

direction in the central beam axis and ignores lateral scatter. Therefore, it does not accurately model the distribution of secondary electrons in heterogeneous media. This limitation causes inaccurate dose calculation in heterogeneous treatment sites such as the lung, bone, or interfaces [15, 16].

2.3.2 Fast Fourier transform (FFT) convolution

This technique reduces computation time greatly because of the invariant kernel assumption for the convolution calculation. Because of this assumption, different kernels at different regions based on the density cannot be used in FFT. Several studies were conducted to circumvent invariant kernel assumptions [17–19]. In 1996, Wong et al. [20] proposed a solution to address problems related to lateral disequilibrium and penumbra in low-density regions because a water kernel was used for entire regions even in low-density regions. The lateral disequilibrium problem was solved by *lateral* scaling of the field size at each depth according to local effective densities to adjust the dose along the central axis in heterogeneities. This technique is based on the ETAR method, by convolving the density at the intersection site with the primary kernel for water. The resultant dose distribution is then inverse scaled according to the effective density to correct the penumbra problem which accounts for the electron transport near the field edge inside a low-density medium with or without lateral disequilibrium.

2.4 Category.4

2.4.1 Convolution-superposition algorithms

The convolution-superposition algorithm is also a model-based algorithm and has two essential parts: 1) TERMA and 2) dose spread kernel. TERMA was first introduced by Ahnesjo et al. in 1987 [21] which is analogous to the Kerma, (the kinetic energy released in medium) and has the same unit as dose. The formula for the TERMA element (T) of the convolution method is given by the following equation

$$T(r') = \frac{\mu}{\rho}\left(\vec{r}',E\right).\Psi\left(\vec{r}'\right) \tag{7}$$

where μ/ρ is the mass attenuation coefficient and Ψ is the primary energy fluence. Then the convolution-superposition is the integration of the TERMA distribution times EDK over the entire volume. EDK is spatially variant and is deformed based on the local density environment to consider interface effects in regions of different densities. Also, to get a more accurate model of the scattering conditions, the kernels must be adjusted according to their direction and orientation at the site of interaction [22].

This method is widely used in TPS because computers are fast enough to do 3D dose calculations by using electron density data derived from CT images in a reasonable amount of time. According to AAPM report 85 (TG-65), the dose calculation accuracy of TPS algorithms should be within 2%. This goal serves as a useful benchmark to evaluate the capabilities of treatment planning algorithms to calculate the dose.

2.4.2 Anisotropic analytical algorithm

Anisotropic Analytical Algorithm (AAA) (Varian Medical System, Inc) is a kernel-based convolution-superposition method. This algorithm was first designed by Ulmer and Kaissl (2005) [23] in cylindrical coordinate and then improved by Tillikainen in 2008 [24]. The AAA dose calculation model has two main components, the configuration algorithm, and the actual dose calculation algorithm. Its configuration is based on the Monte Carlo simulations to determine basic physical parameters and match them with measured clinical beam data. The dose calculation algorithm utilizes separate models for primary photons, scattered extra-focal photons, and contamination electrons. The lateral distribution is adjusted according to the radiological distance to the calculation point for tissue heterogeneities corrections [16, 25]. For the most part, AAA is a pencil beam convolution-superposition algorithm where the pencil beam is compiled from Monte Carlo calculations and adjusted to fit measurements. In this case, two components need to be considered that contribute to final distributions; 1) longitudinal contribution of the pencil beam which is scaled according to Equivalent Length Path (EPL), and 2) contribution from the lateral extension of the pencil beam which is scaled with the density relative to water in directions normal to the pencil beam [26]. In this way, the changes in lateral transport of energy are modeled when the density varies in the irradiated object. Therefore, unlike the pencil beam algorithms, it can consider inhomogeneity correction on both longitudinal and lateral directions. However, many studies indicate the inability of AAA to accurately calculate doses at interfaces and for high atomic number materials such as bone and have shown that the deviation between AAA and measurements exceeds the goal of TG-65 [27–30].

The advantage of the AAA is its relatively short calculation time and its accuracy is better than the pencil beam convolution (PBC) model [30–32].

2.4.3 Collapsed cone convolution

In 1989 Ahnesjo [33] proposed collapsed cone convolution (CCC) method. The CCC algorithm uses the analytical kernel in polar coordinates represented by a set of cones. In this way, it is assumed that all energy is released into coaxial cones of equal solid angle and, from volume elements on the cone axis is approximated to be rectilinearly transported, attenuated, and deposited in volume elements on that axis [7]. The polyenergetic kernels can be described by

$$h(r,\theta) = \frac{A_\theta e^{-\alpha_\theta r} + B_\theta e^{-b_\theta r}}{r^2} \tag{8}$$

where A_θ, α_θ, B_θ, and b_θ are fitting parameters depending on the scattering angle θ and r is radial distance. The first term mainly describes the primary dose and the second term is the scatter dose fraction.

The advantage of the CCC algorithm over standard convolution algorithms is that it can reduce the computation resources. The computation time for the CCC method in heterogenous media is proportional to MN^3 where M is the number of cones and N is the number of voxels along one side of the calculation volume [16]. Different TPSs use the CCC algorithm such as Pinnacle (Philips Inc., Amsterdam, Netherlands), Oncentra MasterPlan (Nucletron, Inc., Columbia, MD, USA), CMS XiO (Elekta AB, Stockholm, Sweden), RayStation (RaySearch Laboratories AB, Stockholm, Sweden), etc.

2.4.4 Monte Carlo

Monte Carlo (MC) is a principle-based algorithm that almost includes all known physical features for photon interactions inside the patient body. Many MC codes have been developed such as BEAMnrc, GEANT4, MNCP, PENELOPE, and XVMC. All of them have two main steps, first, modeling the linac head with all precise details of the target, component dimensions, geometry, locations, and material composition. The second step uses CT data to get morphological and chemical information in terms of mass density, electron density, and atomic composition, which are all required for accurate dose calculation in the tissue.

The MC has the capability of simulating all interactions, therefore it is expected to be accurate. However, its accuracy depends on correct and detailed geometry information of the linac head and the number of particle histories. This statistical uncertainty is proportional to the inverse square root of the generated event numbers [34, 35]. MC dose calculation is slow and time-consuming, so they are not yet applicable in clinics because the dose may recompute repeatedly during planning to get an optimized plan. A few vendors offer Monte Carlo methods in TPS as calculation options for the final dose calculation once the dose optimization is completed.

2.4.5 Acuros XB

Monte-Carlo (MC) dose calculation algorithm is widely considered as the golden dose calculation technique in radiation therapy; however, the calculation time of this method is still long especially where a greater number of particle histories should be used to reduce statistical noise and/or a high spatial resolution is required. An alternative method to MC is the linear Boltzmann transport equation (LBTE) method which solves LBTE refers to grid-based Boltzmann solver (GBBS). GBBS solves the LBTE through discretizing photon and electron fluences in space, energy, and angle to allow a deterministic solution of the transport of radiation through matter. Its calculation accuracy is comparable to MC, and both are convergent methods because the MC algorithm simulates an infinite number of particles, GBBS discretizes the LBTE variables into infinitely small grids, then the two methods should converge to the real solution. However, MC and GBBS have different sources of error, there is statistical noise due to simulating a finite number of particles in Monte Carlo, while most errors in GBBS methods are systematic and their main source is discretization of the solution variables in space, angle, and energy [36, 37]. An algorithm using this technique is based on Attila (Los Alamos National Laboratory, Los Alamos, NM, and Transpire Inc., Gig Harbor, WA). Attila employs linear discontinuous finite-element spatial differencing on a computational mesh consisting of arbitrary tetrahedral elements. The primary photon fluence is analytically transported through ray tracing, and the discrete ordinates method is used for angular differencing of the scattered fluence. Based on Attila, a dose calculation algorithm for external photon beams has been developed on the same methods and implemented in the Varian Eclipse external beam treatment planning system (Varian Medical Systems, Palo Alto, CA, USA) [38]. This new deterministic radiation transport algorithm is Acuros XB (AXB), and it has been well shown by several studies that the accuracy of dose calculation of AXB is more accurate than AAA and is very similar to MC dose calculations [36–38].

3. Beam Modeling of commercial treatment planning systems

In radiotherapy, the ability of TPS to do accurate dose calculation is important. This capability depends on the algorithm of TPS as discussed before and beam modeling. For beam modeling, several dosimetric parameters (e.g., PDDs, profiles, output factors) and non-dosimetric parameters such as MLC design, flattening filter, wedges etc. must be defined precisely. Then the dose calculation algorithm applies the beam model to the patient body or phantom to calculate the dose. The challenge of the beam model is becoming more and more crucial due to advanced treatment techniques such as IMRT and VMAT. In these treatment techniques, each beam consists of multiple segments or control points that are shaped with MLC. Using multiple control points provides this opportunity to deliver conformal dose to the target, however, delivering dose through small segments arises a challenge to accurately calculate the dose due to the complexity of MLC modeling in TPS. Many studies indicate the importance of accurate MLC modeling in TPS for IMRT. In 1998, LoSasso et al. [39] showed an MLC error gap of 1 mm may result 10% error in dose calculation in the sliding window IMRT technique. Cadman et al. [40] reported 12% discrepancy between calculation and measurement due to MLC leaf gap error in step-and-shoot IMRT. Because different commercial TPS have their own features for beam modeling, many guidelines have been published regarding TPS commissioning for IMRT [41, 42]. For example, TG-119 [43] based on the IMRT QA results of five institutions for a set of test cases provides a reference baseline for the accuracy of IMRT commissioning.

In Eclipse, leaf transmission factors and dosimetric leaf gaps (DLGs) are required to model the MLC. The DLG is a beam configuration parameter used to model the effects of rounded MLC leaf ends. Many research papers indicate the effects of DLG on the accuracy of dose calculation in Eclipse TPS [44–47].

In RayStation, modeling of MLC is different from other commercial TPS. The MLC model requires four parameters: leaf-tip offset, leaf-tip width, average transmission factor, and tongue and groove. The leaf-tip width is used for the MLC leaf-end transmission modeling instead of using dosimetric leaf gap (DLG) or rounded leaf-tip radius, and the MLC leaf radiation transmission is modeled using average transmission factor instead of intra-leaf and inter-leaf transmission [48, 49]. According to Chen et al. tongue-and-groove has a minimal effect on IMRT dose calculation, but transmission plays a significant role in this commercial TPS [49].

4. Measurement methods for pre-treatment verification

The process of patient-specific QA usually involves applying an optimized plan using the same beam parameters as those of the patient plan and delivered in the phantom. This process can be done in a number of different ways but according to TG-218 [50], there are three common methods for performing pre-treatment QA. 1) True Composite (TC), 2) Perpendicular field-by-field (PFF), and 3) Perpendicular composite (PC).

4.1 True composite

In this method, phantom or measurement device is placed on the treatment couch and treatment plan is delivered using actual parameters such as MUs, couch, gantry, collimator angles, MLCs, and jaws positions. The phantom or measurement device has been used to integrate dose from all beams of a plan which result in a single dose image for comparison, therefore, this method is a comparison of planned dose vs. measured dose.

4.2 Perpendicular field-by-field

The gantry is fixed at zero degree and the collimator is fixed at the nominal angle in the PFF technique. Therefore, beams are always perpendicular to the phantom surface and are comparing the dose of each beam with each measured beam dose.

4.3 Perpendicular composite

This method is similar to the PFF method, but this is not a comparison of field-by-field. This is the integration dose of all perpendicular field which result in one dose image for analysis.

5. Gamma index

For the purpose of dose comparisons between calculated and measured dose gamma index have been used. Low et al. [51] developed a gamma index (γ) for the quantitative evaluation of dose distributions. This index checks dose difference and distance-to-agreement (DTA) simultaneously in a space that also includes dose, and provides quantitative value which indicates disagreement in the regions that fail the acceptance criteria. A γ comparison is performed between two dose maps: one distribution is the 'reference dose distribution' and the other is the 'evaluated dose distribution'. The ***reference dose distribution*** is referred to as true distribution so it is usually measured data using devices such as ion chamber, film, diode array detector etc., and the ***evaluated dose distribution*** is analyzed for its agreement with the reference and can be the predicted TPS dose distribution. To avoid any confusion, low replaced reference and evaluated terms by measured and calculated respectively. The gamma index calculation is based on Eq. (9):

$$\Gamma(r_R, r_E) = \sqrt{\frac{\Delta r^2(r_R, r_E)}{\delta r^2} + \frac{\Delta D^2(r_R, r_E)}{\delta D^2}} \tag{9}$$

where r_R and r_E are reference points and evaluated point respectively, δr is distance difference criterion and δD is the dose difference criterion. ΔD is dose difference which is calculated using Eq. (10):

$$\Delta D(r_R, r_E) = D_E(r_E) - D_R(r_R) \tag{10}$$

D_E and D_R are the doses at a point in evaluated dose distribution and reference dose distribution respectively.

The γ is the minimum value calculated overall evaluated points:

$$\gamma(r_R) = \min\{\Gamma(r_R, r_E)\}\forall\{r_E\} \tag{11}$$

Regions where γ is less than or equal to 1 corresponds to locations where the calculation meets the acceptance criteria. According to TG-218, criteria for tolerance limit is 2 mm/3% with 95% passing rate [50].

There are two types of gamma calculation which depends on how the percent dose difference (%Diff) is normalized: 1) local normalization method which %Diff is normalized to the doses at each evaluated point, 2) global normalization method which %Diff is normalized usually to the maximum dose within the reference dose distribution. Each method has its own advantages and disadvantages. For example, local gamma will exaggerate %Diff and highlighted failures in low dose regions because in low dose regions the percent dose difference between calculated and measured may exhibit a very large value which results in more failings points. However, in the global method, the dose discrepancies in the low-dose regions could be underestimated which results in a higher passing rate than the local method [52, 53].

5.1 Effect of planned grid size on gamma passing rate

Low et al. [51] presented a powerful tool for dose distribution comparisons in a continuous environment; however, clinical comparisons are usually made between two dose distributions which are sampled at different spatial resolutions. The importance of spatial resolution was first analyzed by Depuydt *et al* in 2001 [54]. They indicated that the pixel size of the compared image needs to be small with respect to acceptance criteria and showed that large grid spacing in the discrete dose distribution, especially in high dose gradient regions causes overestimation of gamma values. Several investigators introduced different solutions to resolve this issue [54–56]. For example, Low and Dempsey [57] showed that by decreasing grid size to 1×1 mm^2, the error in γ reduced to less than 0.2 even in high dose gradient areas. Furthermore, Schreiner et al. [58] reported changing the resolution of the evaluated distribution (from 2.5 mm to 0.24 mm) increase the pass rate from 80.9% to 91.3%. These results are attributed to the behavior of gamma search. When the pixel size of the evaluated distribution is large compared to the reference distribution, many reference pixels would be far away from the nearest evaluated pixel which results in more failing points. Thus, the γ value for many reference pixels reflects significant spatial misalignment purely as an artifact of the coarse evaluated resolution. When the resolution of the evaluated distribution is increased to match that of the reference distribution, this spatial artifact is eliminated because each reference point has a directly corresponding pixel in the evaluated distribution. Increasing the evaluated resolution also provides each reference point with a greater range of dose values for comparison. Based on TG-218 [50], there is a rule of thumb that the resolution of the evaluated (calculated) should be no greater than 1/3 of the DTA and the straightforward solution for reducing artifact in gamma calculation is interpolation when planned grid size is greater than 1 mm (for DTA =3 mm).

6. Dosimetry equipment for pre-treatment verification

Modern radiotherapy techniques like IMRT and VMAT are highly complex modalities due to MLCs motions, gantry rotation, dose rate variation during beam delivery. The advantages of using these techniques are delivery of conformal radiation dose to the target while sparing the surrounding normal tissues and organs-at-risk (OAR) are significantly higher compared to conventional 3D techniques. However, due to the high degree of complexity of these techniques, it is strongly recommended to do pre-treatment verification before dose delivery. For this reason, different types of 2D or 3D detectors such as diode arrays, ionization chambers, film (e.g., Gafchromic film EBT3), electronic portal imaging device (EPID), etc. have been used to ensure that the prescribed treatment dose is delivered within the clinically acceptable error tolerances. Regardless of the type of detector, all of this equipment has spatial limitations because of the discrete placement and physical separation of each detector which may affect GPR results [59, 60].

6.1 Effect of detector resolution on gamma passing rate

As it was mentioned before, phantoms or dosimeter devices used for performing patient-specific QA present spatial resolution limitations which may affect GPR results. Several research has been conducted to show the discrepancy of GPR within different phantoms with different spatial resolutions. Bruschi et al. [59] studied the effect of detector resolution on GPR. Three detectors (PTW OCTAVIUS 4D 729, 1500, and 100 SRS) used in five configurations with different resolutions were utilized in their study. This study indicates the detector resolution can significantly affect the SBRT pre-treatment verification results and a detector with high spatial resolution would be able to detect any kind of error such as those caused by MLC position, collimator, and gantry rotations, etc. In 2017 Woon et al. [61] worked on a similar subject and used three detectors with different resolutions (MapCHECK2, ArcCHECK, and EPID). They demonstrated that MLC errors of greater than 0.5 mm were not distinguishable in measured doses by the MapCheck2/ArcCHECK due to the inferior resolution caused by the large diode spacing relative to the resolution of the EPID. Bailey et al. [53] reported that detector arrays with low-spatial resolution may potentially affect the gamma index analysis by under sampling data. On the other hand, Steers et al. [62] indicated that different detectors show different error sensitivity which depends on the induced type of error and the GPR does not highly depend on detector spatial sampling. Moreover, they showed that increasing spatial sampling not only increase the GPR but also reduces error sensitivity in many cases. This is observed if the increase in the number of sampling results in a higher number of low dose points in the comparison than high dose points, an effect which is increasingly important for globally normalized gamma comparisons [62]. Salari et al. [63] also compared standard density vs. high density measurements of ArcCHECK phantom in Intensity Modulation Radiosurgery (IMRS) cases and compared the GPR values. As shown in **Figure 1**, the results of standard density mode had better GPR for each energy and planned dose grid which is also in good agreement with Steer et al. result. Note that 1 mm and 2 mm represent GS; 6 FFF and 10 FFF for 6 MV FFF and 10 MV FFF beam energies, respectively.

Hussein et al. [64] also conducted research on five commercial QA devices and analyzed the effect of detector resolution on γ. They concluded that different combinations of QA devices and software exhibit varying level of agreement for the same passing rate.

7. Modulation indices

Modern treatment techniques, such as IMRT and VMAT, have enabled the escalation of target dose with fewer side effects to the surrounding OARs by modulation of the treatment plans to achieve the desired dose distribution. In IMRT, MLCs are moving during treatment, thereby delivering a radiation field with a non-uniform intensity while in VMAT technique, in addition to MLC motions, gantry speed and dose rate are also variable when the radiation beam is continuously on. For patients' protection and safety, pretreatment dosimetric verification is done to provide

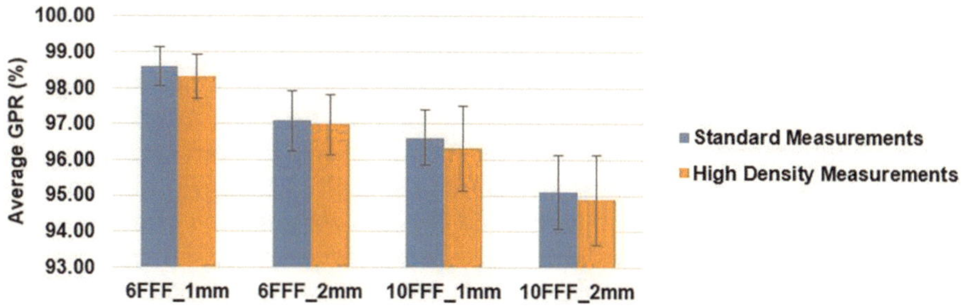

Figure 1.
Comparisons of standard density vs. high-density modes between different planned grid sizes and energy. (reprinted from Salari, et al., "evaluation of parameters affecting gamma passing rate in patient-specific QA's for multiple brain lesions IMRS treatments using Ray-Station treatment planning system. In print: J Appl Clin med Phys. 2021).

sufficient data on the safety and reliability of treatment plans and delivery, even though performing pretreatment dosimetric verification is considered an additional workload. Therefore, a retrospective analysis of which parameter (leaf travel, beam aperture and shapes, control point angular separation, dose rate, and gantry variations) can affect the ability of the TPS to calculate a dose may provide important information on the limits of TPSs for IMRT/VMAT plans. The difference between calculated and measured dose distribution may be affected by the accuracy of the TPS calculation and the delivery accuracy. Discriminating between the two causes of errors is not an easy task. Furthermore, the delivery accuracy of IMRT/VMAT plans can be predicted by the score of plan modulation complexity [65]. For this purpose, many authors introduced or evaluated different Modulation Indices (MI)/parameters to find a correlation between plan complexity and GPR.

Nicolini et al. [66] studied the effect of gantry speed (deg/s) and dose rate (MU/min) on the quality of VMAT plans and showed using a higher dose rate improves plan quality and reduces delivery time. They also used dynamic log files generated by linac controllers to evaluate the delivery accuracy of plans and found out accuracy slightly improved in delivery when using a low dose rate. Wu et al. [67] analyzed the results of dose verification of 924 patients including the relationship between gamma pass rates and the location of lesions, the total number of monitor units, and the maximum area of the collective dose. They observed a correlation between the treatment site and GPR plus a strong negative correlation between total MUs and GPR that indicates increasing MU results in lower GPR. Moreover, a weak negative correlation between the largest area of the acquisition dose and GPR was reported [67]. McNiven et al. [68] proposed *Modulation Complexity Score (MCS)* for step-and-shoot IMRT. This score is contribution of variability in the shape of segments and variations in their area. The range of MCS is from 0 to 1. The lower value of the MCS means higher complexity. This metric provides more information about the plan quality than simple metrics such as total MUs and number of segments, but no correlation was observed between GPR and MCS which is in a good agreement with other research [69, 70]. This index was later adapted by Masi et al. [65] for VMAT plans by substituting control points for segments and called it (MCSv). Also, Masi et al. introduced *Leaf Travel (LT)* as the average distance that MLC is traveling over one arc in VMAT and LTMCS index which takes into account both LT and MCSv and has a range between 0 and 1. Zero shows a higher degree of modulation and leaf motion. They reported a moderate correlation between LT, MCSv, LTMCS, and GPR and a weak correlation between MU and GPR. Hernandez et al. [71] modified LT for multiple arcs or partial arc by dividing LT over arc length (LT/AL). Another index is *Edge Metric (EM)* which was defined by Young et al. [72] and it calculates the complexity as the ratio of MLC side length edge to aperture area. The larger EM index indicates the difference between the positions of adjacent leaves are larger which is closely related to the tongue-and-groove effect. Du et al. in 2014 [73] introduced several MIs to evaluate plan complexity such as *plan averaged beam area (PA), plan averaged beam irregularity (PI), plan averaged beam modulation (PM), and plan normalized MU (PMU).* PA is the average area of beam apertures; PI indicates the non-circularity of the shape of aperture and PM describes to what extent a beam is modulated with multiple smaller apertures. PMU is to compare the total MU among all plans with different prescription dose levels. According to a number of studies [70, 71, 73] MCS, EM, and PI provide similar information. In 2014, Park et al. [74] defined MI_s, $MI_{a,}$ and MI_{total} which MI_{total} unlike previous metrics include both gantry speed and dose rate variations besides MLC motions to quantify the total delivery complexity for VMAT plans. MI_s which evaluate MLC speed was originally introduced by Webb [75] to evaluate the modulation degree of IMRT and were modified by Park et al. for VMAT treatment plans and MI_a evaluates both speed and acceleration of MLCs. They also studied the MCSv and LTMCS and did not see correlations as high as those found in a previous study (Masi et al) to the pre-treatment VMAT QA results.

In summary, various studies were conducted in this area and revealed different results regarding the correlation between plan complexity indices and QA metrics [65–79]. We believe, these differences may depend on the linac model and its commissioning plus TPS limitations such as beam model, dose engine, and algorithm [71, 80, 81].

8. Conclusions

As described in this chapter, there are a number of sources which may contribute and arise different levels of discrepancy between the computed dose by TPS and measurements. Much effort has been devoted to improve the accuracy of dose calculation algorithms, computing technology and measurements, and through all these developments the accuracy of dose calculation and measurements seems close to our clinical goals. Although, the accuracy of dose calculation in homogenous medium (e.g., water) does not much rely on the algorithm, in heterogeneous media such as lungs or bone, the accuracy of calculation depends strongly on the kernels of calculation algorithms and how well they can simulate the actual scattering of photon and electrons. As mentioned previously in this chapter and noted by authors in various literatures, the accuracy of dose calculation algorithms is rated as principle-based algorithms such as Monte Carlo, and the linear Boltzmann transport as the most accurate, followed by model-based algorithms such as CCC, AAA, and PBC in that order for accuracy; and correction-based algorithms. Another important item to be considered is the beam modeling which will directly affect the accuracy of dose calculation where each TPS has its own features to model beams. Therefore, following the beam data measurements, commissioning of the modeled beams becomes a necessary step typically achieved through end-to-end testing. This is to verify dose distribution and accurate computation under different clinical conditions before any clinical use. Moreover, it is important to understand the response and limitations of each equipment used along with gamma index analysis due to different combinations of QA devices and software packages, which may result in varying levels of agreement with the predicted gamma analysis for the same pass-rate criteria. Various reasons result in different correlations between GPR and complexity metrics, hence, these correlations are not generic and should be defined for each TPS.

Conflict of interest

None to report.

Author details

E. Ishmael Parsai* and Elahheh Salari
Department of Radiation Oncology, University of Toledo Medical Center, Toledo, Ohio, USA

*Address all correspondence to: e.parsai@utoledo.edu

References

[1] Khan F, Gerbi B. Treatment planning algorithms: Model-based photon dose calculations. Treatment Planning in Radiation Oncology. 2012;**1(3 ed)**:93-110

[2] Liu HH, Mackie TR, McCullough EC. A dual source photon beam model used in convolution/superposition dose calculations for clinical megavoltage x-ray beams. Medical Physics. 1997;**24**(12): 1960-1974

[3] Cashmore J. The characterization of unflattened photon beams from a 6 MV linear accelerator. Physics in Medicine and Biology. 2008;**53**(7):1933-1946

[4] Zhu XR, Kang Y, Gillin MT. Measurements of in-air output ratios for a linear accelerator with and without the flattening filter. Medical Physics. 2006; **33**(10):3723-3733

[5] Cho W, Kielar KN, Mok E, Xing L, Park JH, Jung WG, et al. Multisource modeling of flattening filter free (FFF) beam and the optimization of model parameters. Medical Physics. 2011;**38**(4): 1931-1942

[6] Papanikolaou N, Battista JJ, Boyer A, Kappas C, Klein E, Mackie TR, et al. Tissue inhomogeneity corrections for megavoltage photon beams. In: Report of Task Group No 65 of the Radiation Therapy Committee of the American Association of Physicists in Medicine. Vernon Boulevard, Madison, USA: Medical Physics. 2004

[7] Papanikolaou N, Stathakis S. Dose-calculation algorithms in the context of inhomogeneity corrections for high energy photon beams. Medical Physics. 2009;**36**(10):4765-4775

[8] Batho HF. Lung corrections in cobalt 60 beam therapy. Journal of the Canadian Association of Radiologists. 1964;**15**:79-83

[9] Sontag MR, Cunningham JR. Corrections to absorbed dose calculations for tissue inhomogeneities. Medical Physics. 1977;**4**(5):431-436

[10] El-Khatib E, Battista JJ. Improved lung dose calculation using tissue-maximum ratios in the Batho correction. Medical Physics. 1984;**11**(3):279-286

[11] Thomas SJ. A modified power-law formula for inhomogeneity corrections in beams of high-energy x rays. Medical Physics. 1991;**18**(4):719-723

[12] Ahnesjo A, Aspradakis MM. Dose calculations for external photon beams in radiotherapy. Physics in Medicine and Biology. 1999;**44**(11):R99-R155

[13] Kappas K, Rosenwald JC. Theoretical and experimental analysis of scatter from inhomogeneous slabs in a 60Co beam: The differential tissue-air ratio method (DTAR). Physics in Medicine and Biology. 1986;**31**(11):1211-1228

[14] Mohan R, Chui C, Lidofsky L. Differential pencil beam dose computation model for photons. Medical Physics. 1986;**13**(1):64-73

[15] Elcim Y, Dirican B, Yavas O. Dosimetric comparison of pencil beam and Monte Carlo algorithms in conformal lung radiotherapy. Journal of Applied Clinical Medical Physics. 2018; **19**(5):616-624

[16] Kim DW, Park K, Kim H, Kim J. History of the photon beam dose calculation algorithm in radiation treatment planning system. Progress in Medical Physics. 2020;**3**(31):54-62

[17] Boyer AL, Mok EC. Calculation of photon dose distributions in an inhomogeneous medium using convolutions. Medical Physics. 1986; **13**(4):503-509

[18] Zhu Y, Boyer AL. X-ray dose computations in heterogeneous media using 3-dimensional FFT convolution. Physics in Medicine and Biology. 1990; **35**:351-368

[19] Wong E, Zhu Y, Van Dyk J. Theoretical developments on fast Fourier transform convolution dose calculations in inhomogeneous media. Medical Physics. 1996;**23**(9):1511-1521

[20] Wong E, Van Dyk J, Zhu Y. Lateral electron transport in FFT photon dose calculations. Medical Physics. 1997; **24**(12):1992-2000

[21] Ahnesjo A, Andreo P, Brahme A. Calculation and application of point spread functions for treatment planning with high energy photon beams. Acta Oncologica. 1987;**26**(1):49-56

[22] Mack A, Weltz D, Scheib SG, Wowra B, Bottcher H, Seifert V. Development of a 3-D convolution/superposition algorithm for precise dose calculation in the skull. Australasian Physical & Engineering Sciences in Medicine. 2006;**29**(1):1-12

[23] Ulmer W, Pyyry J, Kaissl W. A 3D photon superposition/convolution algorithm and its foundation on results of Monte Carlo calculations. Physics in Medicine and Biology. 2005;**50**(8): 1767-1790

[24] Tillikainen L, Helminen H, Torsti T, Siljamaki S, Alakuijala J, Pyyry J, et al. A 3D pencil-beam-based superposition algorithm for photon dose calculation in heterogeneous media. Physics in Medicine and Biology. 2008;**53**(14): 3821-3839

[25] Sievinen J, Ulmer W, Kaissl W. AAA Photon dose calculation model in eclipseTM. Palo Alto: Varian Medical System; 2005

[26] Knoos T, Wieslander E, Cozzi L, Brink C, Fogliata A, Albers D, et al. Comparison of dose calculation algorithms for treatment planning in external photon beam therapy for clinical situations. Physics in Medicine and Biology. 2006;**51**(22):5785-5807

[27] Sterpin E, Tomsej M, De Smedt B, Reynaert N, Vynckier S. Monte Carlo evaluation of the AAA treatment planning algorithm in a heterogeneous multilayer phantom and IMRT clinical treatments for an Elekta SL25 linear accelerator. Medical Physics. 2007;**34**(5): 1665-1677

[28] Robinson D. Inhomogeneity correction and the analytic anisotropic algorithm. Journal of Applied Clinical Medical Physics. 2008;**9**(2):112-122

[29] Ono K, Endo S, Tanaka K, Hoshi M, Hirokawa Y. Dosimetric verification of the anisotropic analytical algorithm in lung equivalent heterogeneities with and without bone equivalent heterogeneities. Medical Physics. 2010;**37**(8):4456-4463

[30] Gete E, Teke T, Kwa W. Evaluation of the AAA treatment planning algorithm for SBRT lung treatment: Comparison with Monte Carlo and Homogeneous pencil beam dose calculations. Journal of Medical Imaging and Radiation Sciences. 2012;**43**(1):26-33

[31] Panettieri V, Barsoum P, Westermark M, Brualla L, Lax I. AAA and PBC calculation accuracy in the surface build-up region in tangential beam treatments. Phantom and breast

case study with the Monte Carlo code PENELOPE. Radiotherapy and Oncology. 2009;**93**(1):94-101

[32] Bragg CM, Conway J. Dosimetric verification of the anisotropic analytical algorithm for radiotherapy treatment planning. Radiotherapy and Oncology. 2006;**81**(3):315-323

[33] Ahnesjo A. Collapsed cone convolution of radiant energy for photon dose calculation in heterogeneous media. Medical Physics. 1989;**16**(4):577-592

[34] De Martino F, Clemente S, Graeff C, Palma G, Cella L. Dose calculation algorithms for external radiation therapy: An overview for practitioners. Applied Sciences. 2021;**11**(15). DOI: 10.3390/app11156806

[35] Jeraj R, Keall P. The effect of statistical uncertainty on inverse treatment planning based on Monte Carlo dose calculation. Physics in Medicine and Biology. 2000;**45**(12): 3601-3613

[36] Vassiliev ON, Wareing TA, McGhee J, Failla G, Salehpour MR, Mourtada F. Validation of a new grid-based Boltzmann equation solver for dose calculation in radiotherapy with photon beams. Physics in Medicine and Biology. 2010;**55**(3): 581-598

[37] Han T, Followill D, Mikell J, Repchak R, Molineu A, Howell R, et al. Dosimetric impact of Acuros XB deterministic radiation transport algorithm for heterogeneous dose calculation in lung cancer. Medical Physics. 2013;**40**(5):051710

[38] Yan C, Combine AG, Bednarz G, Lalonde RJ, Hu B, Dickens K, et al. Clinical implementation and evaluation of the Acuros dose calculation algorithm. Journal of Applied Clinical Medical Physics. 2017;**18**(5):195-209

[39] LoSasso T, Chui CS, Ling CC. Physical and dosimetric aspects of a multileaf collimation system used in the dynamic mode for implementing intensity modulated radiotherapy. Medical Physics. 1998;**25**(10): 1919-1927

[40] Cadman P, Bassalow R, Sidhu NP, Ibbott G, Nelson A. Dosimetric considerations for validation of a sequential IMRT process with a commercial treatment planning system. Physics in Medicine and Biology. 2002; **47**(16):3001-3010

[41] IAEA. TecDoc 1540: Specification and Acceptance Testing of Radiotherapy Treatment Planning Systems. Vienna: IAEA; 2007

[42] IAEA. TecDoc 1583: Commissioning of Radiotherapy Treatment Planning Systems: Testing for Typical External Beam Treatment Techniques. Vienna: IAEA; 2008

[43] Ezzell GA, Burmeister JW, Dogan N, LoSasso TJ, Mechalakos JG, Mihailidis D, et al. IMRT commissioning: Multiple institution planning and dosimetry comparisons, a report from AAPM task group 119. Medical Physics. 2009;**36**(11): 5359-5373

[44] Zhang Y, Le AH, Tian Z, Iqbal Z, Chiu T, Gu X, et al. Modeling Elekta VersaHD using the Varian eclipse treatment planning system for photon beams: A single-institution experience. Journal of Applied Clinical Medical Physics. 2019;**20**(10):33-42

[45] Middlebrook ND, Sutherland B, Kairn T. Optimization of the dosimetric leaf gap for use in planning VMAT

treatments of spine SABR cases. Journal of Applied Clinical Medical Physics. 2017;**18**(4):133-139

[46] Shende R, Patel G. Validation of Dosimetric leaf gap (DLG) prior to its implementation in treatment planning system (TPS): TrueBeam millennium 120 leaf MLC. Reports of Practical Oncology and Radiotherapy. 2017;**22**(6):485-494

[47] Kim J, Han JS, Hsia AT, Li S, Xu Z, Ryu S. Relationship between dosimetric leaf gap and dose calculation errors for high definition multi-leaf collimators in radiotherapy. Physics and Imaging in Radiation Oncology. 2018;**5**: 31-36

[48] Mzenda B, Mugabe KV, Sims R, Godwin G, Loria D. Modeling and dosimetric performance evaluation of the RayStation treatment planning system. Journal of Applied Clinical Medical Physics. 2014;**15**(5):4787

[49] Chen S, Yi BY, Yang X, Xu H, Prado KL, D'Souza WD. Optimizing the MLC model parameters for IMRT in the RayStation treatment planning system. Journal of Applied Clinical Medical Physics. 2015;**16**(5):322-332

[50] Miften M, Olch A, Mihailidis D, Moran J, Pawlicki T, Molineu A, et al. Tolerance limits and methodologies for IMRT measurement-based verification QA: Recommendations of AAPM task group No. 218. Medical Physics. 2018; **45**(4):e53-e83

[51] Low DA, Harms WB, Mutic S, Purdy JA. A technique for the quantitative evaluation of dose distributions. Medical Physics. 1998; **25**(5):656-661

[52] Park JM, Kim JI, Park SY, Oh DH, Kim ST. Reliability of the gamma index analysis as a verification method of volumetric modulated arc therapy plans. Radiation Oncology. 2018;**13**(1):175

[53] Bailey DW, Nelms BE, Attwood K, Kumaraswamy L, Podgorsak MB. Statistical variability and confidence intervals for planar dose QA pass rates. Medical Physics. 2011;**38**(11):6053-6064

[54] Depuydt T, Van Esch A, Huyskens DP. A quantitative evaluation of IMRT dose distributions: Refinement and clinical assessment of the gamma evaluation. Radiotherapy and Oncology. 2002;**62**(3):309-319

[55] Wendling M, Zijp LJ, McDermott LN, Smit EJ, Sonke JJ, Mijnheer BJ, et al. A fast algorithm for gamma evaluation in 3D. Medical Physics. 2007;**34**(5):1647-1654

[56] Ju T, Simpson T, Deasy JO, Low DA. Geometric interpretation of the gamma dose distribution comparison technique: Interpolation-free calculation. Medical Physics. 2008;**35**(3):879-887

[57] Low DA, Dempsey JF. Evaluation of the gamma dose distribution comparison method. Medical Physics. 2003;**30**(9): 2455-2464

[58] Schreiner LJ, Salomons G, Holmes O. Analysis and evaluation of planned and delivered dose distributions: Practical concerns with γ- and χ- evaluations. Journal of Physics Conference Series (Online). 2013;**444**(1):9

[59] Bruschi A, Esposito M, Pini S, Ghirelli A, Zatelli G, Russo S. How the detector resolution affects the clinical significance of SBRT pre-treatment quality assurance results. Physica Medica. 2018;**49**:129-134

[60] Stathakis S, Myers P, Esquivel C, Mavroidis P, Papanikolaou N.

Characterization of a novel 2D array dosimeter for patient-specific quality assurance with volumetric arc therapy. Medical Physics. 2013;**40**(7):071731

[61] Woon W, Ravindran PB, Ekayanake P, SV, Lim YY, Khalid J. A study on the effect of detector resolution on gamma index passing rate for VMAT and IMRT QA. Journal of Applied Clinical Medical Physics. 2018;**19**(2): 230-248

[62] Steers JM, Fraass BA. IMRT QA and gamma comparisons: The impact of detector geometry, spatial sampling, and delivery technique on gamma comparison sensitivity. Medical Physics. 2021;**48**(9):5367-5381

[63] Salari E, Parsai EI, Shvydka D, Sperling N. Evaluation of parameters affecting gamma passing rate in patient-specific QA's for multiple brain lesions IMRS treatments using Ray-Station treatment planning system. Journal of Applied Clinical Medical Physics. 2021;**23**(1):e13467. DOI: 10.1002/acm2.13467

[64] Hussein M, Rowshanfarzad P, Ebert MA, Nisbet A, Clark CH. A comparison of the gamma index analysis in various commercial IMRT/VMAT QA systems. Radiotherapy and Oncology. 2013;**109**(3):370-376

[65] Masi L, Doro R, Favuzza V, Cipressi S, Livi L. Impact of plan parameters on the dosimetric accuracy of volumetric modulated arc therapy. Medical Physics. 2013;**40**(7): 071718

[66] Nicolini G, Clivio A, Cozzi L, Fogliata A, Vanetti E. On the impact of dose rate variation upon RapidArc implementation of volumetric modulated are therapy. Medical Physics. 2011;**38**(1):264-271

[67] Wu S, Chen J, Li Z, Qiu Q, Wang X, Li C, et al. Analysis of dose verification results for 924intensity-modulated radiation therapy plans. Precision Radiation Oncology. 2018;**2**:125-130

[68] McNiven AL, Sharpe MB, Purdie TG. A new metric for assessing IMRT modulation complexity and plan deliverability. Medical Physics. 2010; **37**(2):505-515

[69] Rajasekaran D, Jeevanandam P, Sukumar P, Ranganathan A, Johnjothi S, Nagarajan V. A study on the correlation between plan complexity and gamma index analysis in patient specific quality assurance of volumetric modulated arc therapy. Reports of Practical Oncology and Radiotherapy. 2015;**20**(1):57-65

[70] Glenn MC, Hernandez V, Saez J, Followill DS, Howell RM, Pollard-Larkin JM, et al. Treatment plan complexity does not predict IROC Houston anthropomorphic head and neck phantom performance. Physics in Medicine and Biology. 2018;**63**(20): 205015

[71] Hernandez V, Saez J, Pasler M, Jurado-Bruggeman D, Jornet N. Comparison of complexity metrics for multi-institutional evaluations of treatment plans in radiotherapy. Physics and Imaging in Radiation Oncology. 2018;**5**:37-43

[72] Younge KC, Matuszak MM, Moran JM, McShan DL, Fraass BA, Roberts DA. Penalization of aperture complexity in inversely planned volumetric modulated arc therapy. Medical Physics. 2012;**39**(11):7160-7170

[73] Du W, Cho SH, Zhang X, Hoffman KE, Kudchadker RJ. Quantification of beam complexity in intensity-modulated radiation therapy

treatment plans. Medical Physics. 2014; **41**(2):021716

[74] Park JM, Park SY, Kim H, Kim JH, Carlson J, Ye SJ. Modulation indices for volumetric modulated arc therapy. Physics in Medicine and Biology. 2014; **59**(23):7315-7340

[75] Webb S. Use of a quantitative index of beam modulation to characterize dose conformality: Illustration by a comparison of full beamlet IMRT, few-segment IMRT (fsIMRT) and conformal unmodulated radiotherapy. Physics in Medicine and Biology. 2003;**48**(14): 2051-2062

[76] Crowe SB, Kairn T, Kenny J, Knight RT, Hill B, Langton CM, et al. Treatment plan complexity metrics for predicting IMRT pre-treatment quality assurance results. Australasian Physical & Engineering Sciences in Medicine. 2014;**37**(3):475-482

[77] Heilemann G, Poppe B, Laub W. On the sensitivity of common gamma-index evaluation methods to MLC misalignments in Rapidarc quality assurance. Medical Physics. 2013;**40**(3): 031702

[78] Nauta M, Villarreal-Barajas JE, Tambasco M. Fractal analysis for assessing the level of modulation of IMRT fields. Medical Physics. 2011; **38**(10):5385-5393

[79] Wu S, Chen J, Li Z, Qiu Q, Wang X, Li C, et al. Analysis of dose verification results for 924 intensity-modulated radiation therapy plans. Precision Radiation Oncology. 2018; **2**(4):125-130

[80] Hernandez V, Abella R, Calvo JF, Jurado-Bruggemann D, Sancho I, Carrasco P. Determination of the optimal tolerance for MLC positioning in sliding window and VMAT techniques. Medical Physics. 2015;**42**(4):1911-1916

[81] Van Esch A, Huyskens DP, Behrens CF, Samsoe E, Sjolin M, Bjelkengren U, et al. Implementing RapidArc into clinical routine: A comprehensive program from machine QA to TPS validation and patient QA. Medical Physics. 2011;**38**(9):5146-5166

Chapter 4

Absolute, Reference, and Relative Dosimetry in Radiotherapy

Carlos Eduardo de Almeida and Camila Salata

Abstract

The correct estimation of the dose to be delivered to a patient is strongly dependent on a correct dosimetry procedure. To achieve this purpose, it must ensure that the measurement results represent the best possible value reported with its typical uncertainties, and the results must be comparable with other institutions. The International Metrological Network fundamentally seeks to standardize the processes and the methodologies among the various laboratories in the world. The concept and structure of the various levels of laboratories can be defined as primary or secondary standards laboratories. The absolute dosimetry refers to the measurement of a quantity with an instrument of the highest metrological quality, which allows its determination in accordance with its definition, usually carried out in Primary Laboratories. The main quantities of interest for the absolute dosimetry are exposure, air kerma, and absorbed dose to air and to water. This entire chain of measurements and formalism must consider the specific physical conditions of the interaction processes between the radiation beam with the detector in the measurement processes, aiming to ensure the least possible uncertainty in the dose delivered to the patient.

Keywords: air kerma, absorbed dose to water, primary standards, absolute dosimetry, traceability

1. Introduction

A dosimetric procedure aims to estimate a quantity to guarantee the delivery of the correct prescribed dose to a patient or the dose resulting from a diagnostic procedure.

To achieve this purpose, one must ensure that:

- the measurement results represent the best possible value reported with its typical uncertainties, using the appropriate calibration coefficients and the correction factors necessary to adjust the measured value to the true value;

- the measurement results made by different institutions must be comparable when performed under similar reference conditions such as those established by international protocols, i.e., TRS#398-IAEA [1] TRS # 469 [2];

- the clinical results of different institutions can be comparable if the air Kerma or absorbed dose, in addition to the biological clinical parameters, is well known and fully described.

To fulfill those premises, the radiation detectors must be calibrated following a universal protocol agreed among the professional societies, and the quantities referenced to the standards at the BIPM as it was decided by the Metro Convention. The dissemination of these quantities until the final user is done through the calibration laboratory in each country either national or secondary following a logical chain of events as described in **Figure 1**.

At this point, one must highlight the importance of a network called: International Metrological Network, which fundamentally seeks to standardize the processes and the methodologies among the various laboratories in the world. As a result, the measurement of the main quantities of interest in radiotherapy, radiology, and radioprotection such as air Kerma and absorbed dose to water allows the clinical results and the biological effects to be compared scientifically among different users, with an acceptable level of uncertainties for each area.

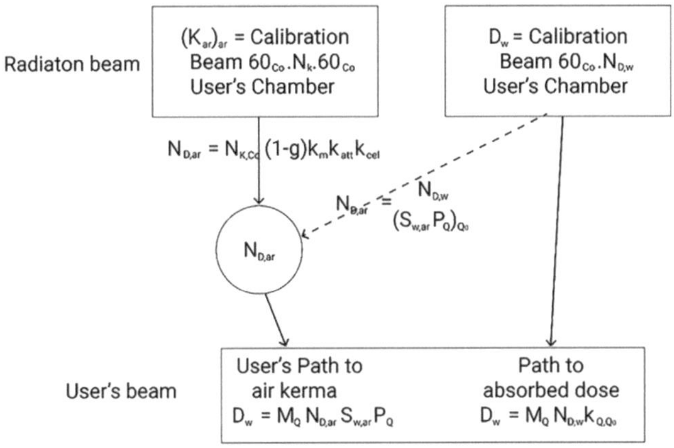

Figure 1.
The main steps involved in the implementation of the quantities: (a) air kerma and (b) absorbed dose to water.

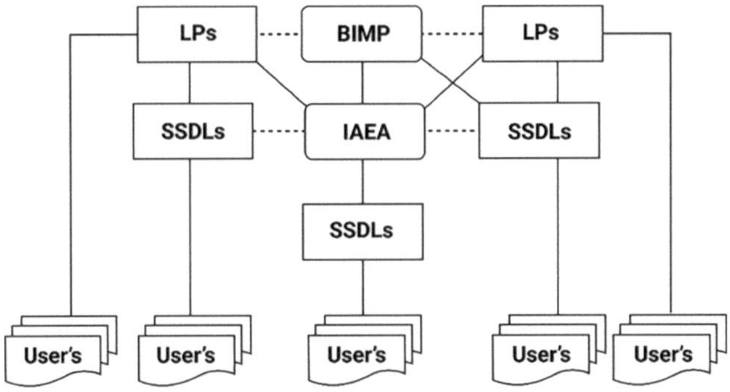

Figure 2.
The international network of ionizing radiation metrology showing the traceability process between the primary standards coordinated by the BIPM followed by the network of the secondary laboratories traceable to the IAEA Laboratories, the BIPM or any other primary laboratory and the final user [1, 2].

The two steps shown above in **Figure 1** constitute a simplification of the various levels of complexity that represent the metrological chain, and the algorithms used now are best illustrated now in **Figure 2**.

2. Standards laboratories

The concept and structure of the various levels of laboratories can be defined as:

2.1 Primary standards dosimetry laboratory (PSDL)

Location where instruments with the highest metrological quality are used, the quantities are measured according to their definition, that is, in an absolute way. To reach this level, very sophisticated equipment, computer control systems, experimental arrangement, and very skilled staff are required, resulting in very small uncertainties, results impossible to be reproduced at the end user's environment.

Those laboratories use free air chambers for air kerma standards in the low and medium energy X-ray beams; water or graphite calorimeters for absorbed dose standard to water or graphite; Fricke dosimeter is a standard for absorbed dose to water and ionization chambers with a well-known volume as standard for either air Kerma for gamma ray beams emitted by a collimated ^{60}Co or absorbed dose to graphite using a large variety of photons and electron beams.

To carry out periodical comparisons involving all National laboratories to ensure the appropriate metrological consistency within the metrological network in a decision agreed by tall country's signatories of the Metro Convention, the BIPM was designated to carry out this task, as shown in **Figure 3**.

2.2 Secondary standards dosimetry laboratory (SSDL)

Location where high-quality metrological instruments are used, though its calibration by one of the PSDL is required to assure that the users' instruments are traceable to the national and international metrological network. In some situations, the IAEA-SSDL Laboratory provides periodical calibration to the members of the IAEA-SSDL network, and QA auditing is also conducted.

The SSDL are recognized and accredited by the country's metrological authority such as the National Laboratory, as it is responsible for disseminating the quantities to the final user in their country ensuring the proper metrological coherence among users with reference to their standards [1, 2]. Since it is possible to find more than one SSDL in one country, an internal network must be established, and periodic comparison must be carried out by the National Laboratory.

In this way, users of ionizing radiation sources will be tracked to the National and International Network with their intercomparable results.

Tips:

- It is not forbidden that the user calibrates their instruments in a PSDL outside the country instead of their SSDL. The drawback is the calibration cost in addition to transportation, insurance, customs clearance expenses, which makes this option too onerous and objectively unnecessary;

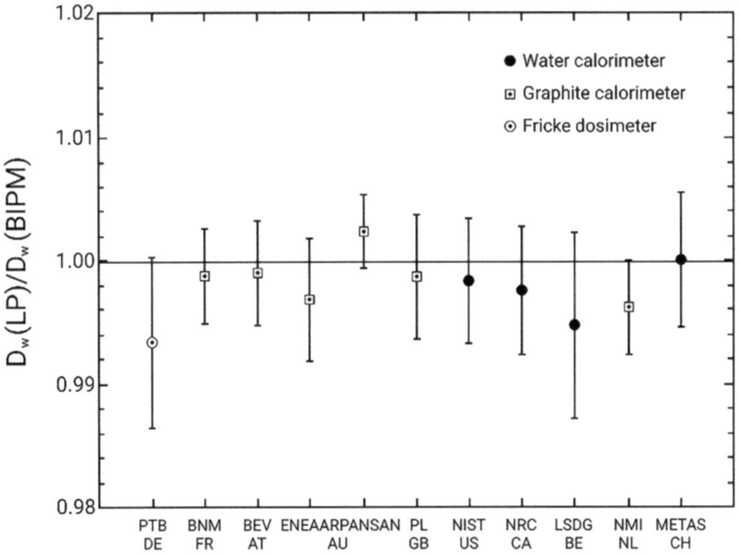

Figure 3.
Typical example of the result of one of the comparisons conducted by BIPM with several national laboratories for the quantity of absorbed dose to water using three different methods: Water calorimeter, graphite calorimeter, and the Fricke system [3].

- Carrying out calibrations in the country's laboratories reinforces the metrological consistency between users and the national laboratory.

2.3 Users level

Location where the calibration procedures of diagnostic and treatment machines are carried out under conditions such as those in which the instruments were calibrated. When using the formalism, for example, from the TRS#398 [1] or similar, it is essential that the measurement systems were calibrated in a laboratory traced to the metrological network.

In this situation, the instruments used can be classified as:

a. reference instrument (the one with the highest level of metrological quality in the institution);

b. field instrument (instrument used in daily routine that can be equal to the first one). This is recommended since some legislations require two sets, one of which could be the reference.

If the institution has only one treatment machine, it is recommended to leave a fixed dosimetry set on the control room bench with the cables passed through the wall of the treatment room, avoiding passing the cable under the door risking damaging it, and the other set as the institutional reference. If you have two treatment machines, leave each system fixed on each machine and as part of the periodic QA program, perform cross-calibration changing the electrometers and performing the measurements. If the values differ consistently by more than 1% between them, use another calibrated chamber on both machines.

The stability test of the dosimetry system shall be performed every three months with a source of ^{90}Sr or ^{137}Cs, as required by the regulatory authority. This test is accepted as a good indicator of the performance of the measurement set, which must include the leakage, repeatability, and linearity tests.

If the QA documentation demonstrates the stability of your system in other ways, it may also be accepted.

Since the numerical values of the uncertainties increase as we go down in the metrological chain, there is a demand for a high-quality measuring system, careful instrument handling procedures especially for the cables and connectors, instrument warm-up, proper documentation, and finally a consistency in positioning the experimental setup.

Measurement systems (ion chamber, electrometer, and cable) must be calibrated when purchased, unless they are calibrated by the manufacturer if it has an accredited laboratory, when they undergo any repairs, and every 2 years regardless of any problem. The calibration coefficient is given for the quantity of absorbed dose to water at the reference conditions. This coefficient is directly traceable to the national and international metrology network. It may be possible to calibrate the ion chamber separately from the electrometer and then use the chambers with different electrometers or vice versa.

3. Absolute, reference, and relative dosimetry

In general, there is a certain conceptual confusion not only by the users but also by the manufactures when using the concepts of absolute dosimetry, reference dosimetry, and relative dosimetry. Andreo et al. [3] very clearly discuss the differences between the three concepts so that they can be used properly.

3.1 Absolute dosimetry

It refers to the measurement of a quantity with an instrument of the highest metrological quality, which allows its determination in accordance with its definition. In general, it is carried out in Primary Laboratories.

For example, the quantity Exposure, X, as defined by ICRU 33 [4], is the result of the quotient of dQ and dm, where dQ is the absolute value of the charge produced by ions of the same sign in the mass of air, when electrons (négatron or positrons), released by photons in an air mass dm, are completely stopped in the air. The unit for the SI system is C/kg, but its special unit is the long-used Roentgen, equaling 2.58.10 C^{-4} C kg^{-1}.

Measures of the quantity Exposure, because of the air Kerma, are of great importance as they constitute the stakes of the metrological chain. They are directly related to the absorbed dose calibrations of the high energy photon and electron beams used in radiotherapy, radiobiology studies, and radioprotection measurements; the latter for the moment entirely dependent on the quantity air Kerma.

3.2 Formalism for the absolute determination of exposure, air kerma, and absorbed dose to water quantities from experimental measurements

3.2.1 Determination of the exposure

The determination of the exposure can be obtained through two methods, both with an ionization chamber:

Method 1. Free air chamber.

Unlike wall chambers, free air chambers do not have walls, so the interaction process occurs within the air volume defined by the electric field defined between the guard ring and the collector plate inside the chamber, to obtain the electronic equilibrium. The thickness of the air layer varies depending on the energy fluence of the beam, and for this reason, two chambers with different volumes are used for energies up to 150 kVp and 300 kVp, respectively. A typical diagram of a free air chamber is illustrated in **Figure 4**.

This process is more largely described by [5], where the formalism for estimating the quantity air Kerma, including typical correction factors, is described in the Eq. (1):

$$K_{air} = \frac{Q_{air}}{\rho} \bullet V \bullet \frac{w}{e} \bullet \frac{1}{1-g} \bullet K_{att} \bullet K_{sc} \bullet k_e \tag{1}$$

Where:

K_{att} = attenuation of the primary beam in air column between the diaphragm and the collector volume;

K_{sc} = additional ionization collected caused by the scattering inside the chamber,

k_e= ionization lost by the shock of the electrons with the electrode;

$\frac{w}{e}$ = average energy needed to produce a pair of ions;

g = the fraction of energy lost by the bremsstrahlung effect;

ρ = air density under the measurement conditions, considering the air compressibility factor that corrects its deviation from the perfect gas law;

V = sensitive volume of the chamber in which charges are produced and collected;

Q_{air} = is the charge produced in the air mass defined as the sensitive volume v of the chamber;

Method 2. Cavity chamber.

This method uses a cavity chamber, with a known volume, with the formalism proposed by [6] and extended by [7]. One must consider the cavity dimensions, the presence of the wall and a central electrode, in addition to the various correction

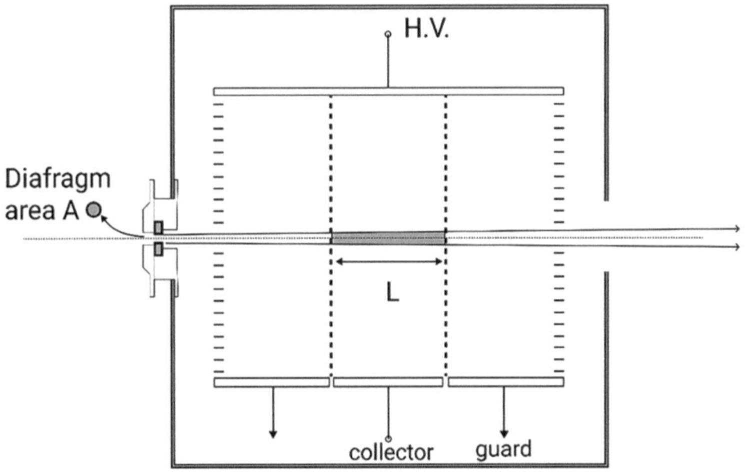

Figure 4.
Typical diagram of a free air chamber where several important components can be identified, such as the diaphragm or frontal collimator with an area a, the collector electrode, and the guard plates when subjected to the same collector potential define the sensitive volume of the chamber.

factors empirically derived such as environmental quantities and measurement statistics. The characteristics of a chamber of this type used in several primary laboratories are described in **Figure 5**.

The final volume measured in the chamber described in **Figure 5** is 1.076 ± 0.003 cm^3, and the graphite caps are used to determine the wall attenuation using the extrapolation method. The graphite complements are added to the base of the chamber after the insertion of each cap to preserve the spatial conditions of scattering. Recently, the wall attenuation value was recalculated by [8] using the Monte Carlo technique, whose result, though slightly different than the experimental one, is more accurate and with less uncertainty.

The primary Standard shown in **Figure 5** is a cylindrical graphite chamber built by the Austrian National Laboratory, with its volume defined by the same laboratory, constructed of ultra-pure graphite (99.99%) with an excellent insulating system to minimize the "leakage" and the polarization effects, guaranteeing an excellent long-term stability and a metrological quality compatible with similar standards, as reported by [9–11].

Its sensitive volume was estimated by the Ostereich Forschung Centrum and reported by [12] from the internal physical dimensions of the chamber, defined with an uncertainty of 0.1% after subtracting the electrode volume according to **Figure 5**, and including the additional sensitive volume in the electrode base.

Thus, according to the Bragg-Gray principle, the measure of ionization in the center of the chamber in its absence is defined by Eq. (2):

$$X = \frac{I}{\rho} \bullet V \bullet sc, a \bullet \left(\frac{(\mu_{en}/\rho_{air})}{(\mu_{en}/\rho)}\right)_C \bullet \Pi K_j \tag{2}$$

Where:

I = ionization current resulting from the collection of ions produced in the air within the chamber cavity, considering the attenuation of the air between the source and the chamber;

V = sensitive volume of the chamber in which charges are produced and collected;

ρ = density of the air under the measurement conditions, considering the air compressibility factor that corrects its deviation from the perfect gas law;

sc, a = the ratio of the restricted stopping power between graphite and air, calculated based on the Spencer-Attix theory [3] taking into account the average value of the energy in the electron spectrum generated by the Compton effect; considering as cutoff energy of 17.5 keV, the cavity size and the average excitation energy of 78 eV for carbon and 85.7 eV for air;

$\left(\frac{(\mu_{en}/\rho_{air})}{(\mu_{en}/\rho)}\right)_C$ = the ratio of mass-energy absorption coefficients for air and graphite used from the work of Hubbel and Seltzer [13];

ΠK_j = the product of several correction factors:

kl = leakage correction;

kh = correction for the presence of water vapor once exposure X is set to dry air;

kst = correction for scattering on the chamber stem;

krn = correction due to radial beam non-uniformity;

kan = correction due to axial beam non-uniformity;

kw = correction due to attenuation of the wall chamber;

kcep = origin of electron production;

kt,p = mass correction for reference temperature and pressure;

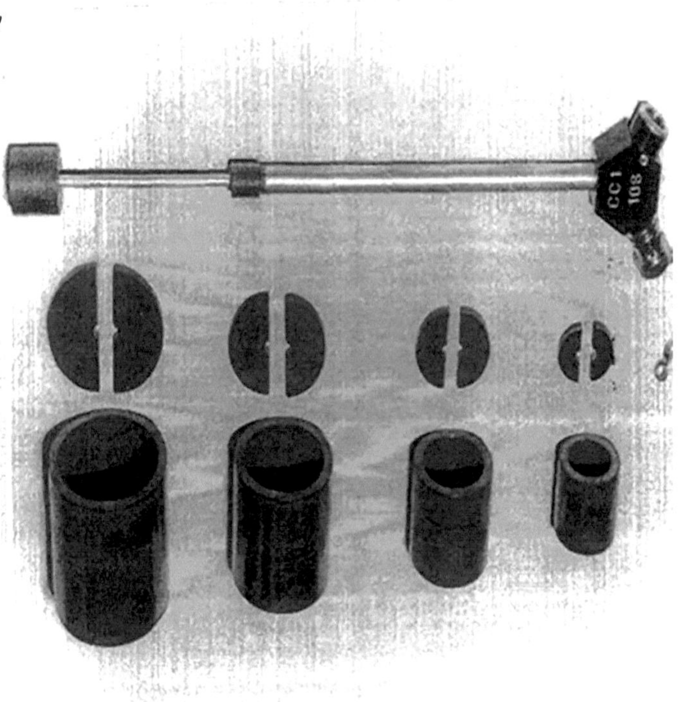

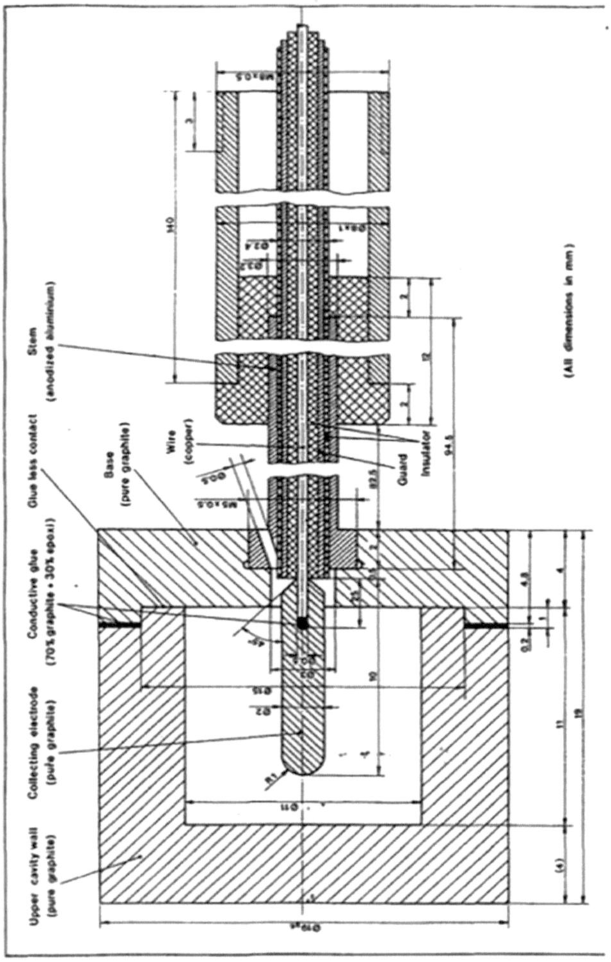

Figure 5.
Image represents the physical diagram, with the internal and external dimensions of the cylindrical chamber.

3.2.2 Determination of the air kerma (K_{air})

The determination of the air kerma (K_{air}) from the measurements of the exposure X follows the formalism below:

$$K_{air} = \frac{X}{1-g} \bullet \frac{w}{e} \tag{3}$$

Where:
X = the air exposure value (X) obtained in accordance with Eq. (2);
g = the fraction of energy lost by the bremsstrahlung effect;
$\frac{w}{e}$ = average energy needed to produce a pair of ions.

3.2.3 Determination of the absorbed dose to air (D_{air})

The determination of the absorbed dose to air (Dair), measured by a standard instrument, is defined as the energy delivered to a mass of air of the well-known sensitive volume of the ionization chamber, defined by the relation:

$$D_{air} = Q_{air} \bullet \frac{W_{air}/e}{m_{air}} \tag{4}$$

Where.
Q_{air} = is the charge produced in the air mass defined as the sensitive volume v of the chamber;
W_{air}/e = average energy needed to produce a pair of ions, its product being equal to the energy given to the air mass m_{air} of the reference sensitive volume;
m_{air} = equal to the product of the air density ρ_{air} and the sensitive volume v.

This measurement may require the use of a set of factors necessary to correlate the reading of the measurement system with the final value of the quantity, such as absorbed dose. The measurements must be carried out under the well-standardized reference conditions, that is: radiation field of 10 x 10 cm^2 on the surface of the phantom, SSD (source surface distance) equal to 100 cm, with the center of the chamber positioned at 5 cm depth, reference temperature of 22°C (reference in Brazil), atmospheric pressure of 101.3 kPa, and relative humidity between 30 and 70% (**Table 1**).

3.2.4 Determination of the absorbed dose to water (D_w)

Method 1. Measurement performed using a graphite or water calorimeter.

Year	$K_{LNMR}I/K_{BIPM}$	u_c	Reference
1986	1.0006	0.0026	de Almeida and Niatel [9]
1996	1.0004	0.0023	Allisy-Roberts et al. [14]
2005	1.0007	0.0022	Allisy-Roberts et al. [15]

uc = combined uncertainty.

Table 1.
Typical history of air kerma standard traceability between two laboratories LNMRI and BIPM.

A Calorimeter measures the quantity absorbed dose to water or to graphite according to its definition, that is, from the increase in temperature in the medium due to a process of radiation induction. This evaluation is done by thermistors installed in the calorimeter body filled with high-purity water, as reported by Malcolm [16]. The calorimeter, in this case, your heart (nucleus), is placed at the reference depth in a 30 cm x 30 cm x 30 cm phantom. The measured signal is generally very low, on the order of 1 mK for an absorbed dose of 2 Gy, and its reproducibility is an important factor. Due to its complexity, it is suitable for use not in clinical settings, but in National Metrology Laboratories or research (**Figure 6**).

An important parameter is the magnitude of the heat defect, that is, the fraction of energy that is not released in the form of heat, being material dependent, this effect being more significant in graphite.

The typical temperature fluctuation obtained when using a radiation source consists of three basic regions:

- the pre-trend that is prior to the irradiation, where fluctuation is stable,

- a constant and almost linear region, when the temperature rises; corresponds to the moment that the source enters the calorimeter being kept in a fixed position, this being the measurement point of the thermistors while the irradiation lasts;

- the post-trend, which is the region that exhibits the behavior of water temperature at time intervals after removal of the source from the calorimeter. The post-trend has a characteristic thermal profile and includes a relative region of low temperature rise that is governed by the increase in temperature gradient created in the water due to direct dose deposition in the water. This can be followed by a sudden increase in temperature due to the decay process of the effect source reaching the measurement point.

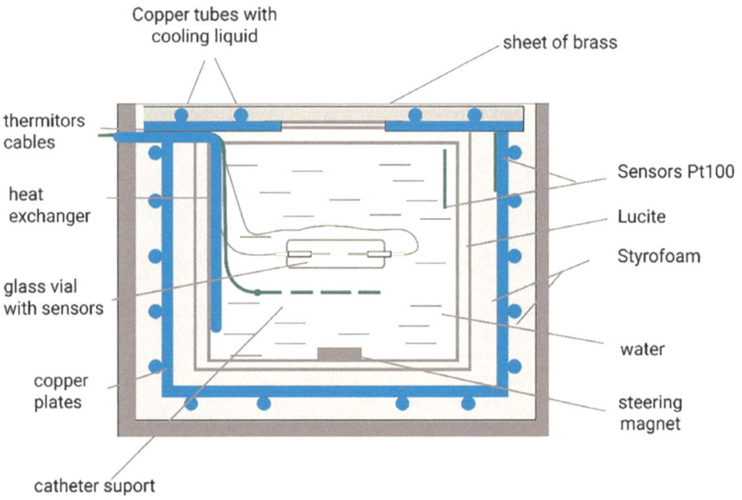

Figure 6.
Shows a schematic diagram of the Domen-type water calorimeter, built jointly with the Canadian McGill University and reported by Rosado and de Almeida [17] to be operated with non-circulating water at 4.0°C.

Using a model of heat conduction in water, the onset time of this sudden temperature rise can be accurately predicted as a function of the distance between the measurement point and the source.

Specifically, for a standard of absorbed dose to water such as the calorimeter, the dose D_w at a point in the water at a given distance (r) from the thermistor corresponds to the measured temperature increase at that point (ΔT) being quantified through the relationship:

$$D_w = \Delta T_w \bullet cw \bullet kt \bullet kc \bullet kv \bullet kdd \bullet k\rho \bullet kHD \tag{5}$$

where:

ΔT_w • = increase in the temperature;

cw = specific heat of the water;

kt = transient effect on the thermistor response due to dose deposition;

kc = conductive transfer of heat due to the excess of heat from the glass components and temperature gradients;

kv = conductive transfer of heat when water temperature is different from 4°C;

kp = disturbance caused in the radiation field due to the presence of the heart (core) of the calorimeter and thermistors, calculated by Monte Carlo simulation;

kdd = refers to the non-uniformity of radiation the beam;

$k\rho$ = variation in the density of water due to the presence of the calorimeter;

kHD = the heat defect, that is, the difference between the absorbed energy and the energy that appears as heat due to chemical reactions induced by radiation.

One of the advantages of the water calorimeter is that the quantity of absorbed dose to water is being measured directly in water, while in the case of using graphite, a graphite to water conversion factor is necessary.

Method 2. Measurement performed on the graphite phantom using a known volume ionization chamber.

In general, the measurement of the absorbed dose to water D_w [1] is carried out under the same reference conditions as mentioned before, as illustrated in **Figure 7**.

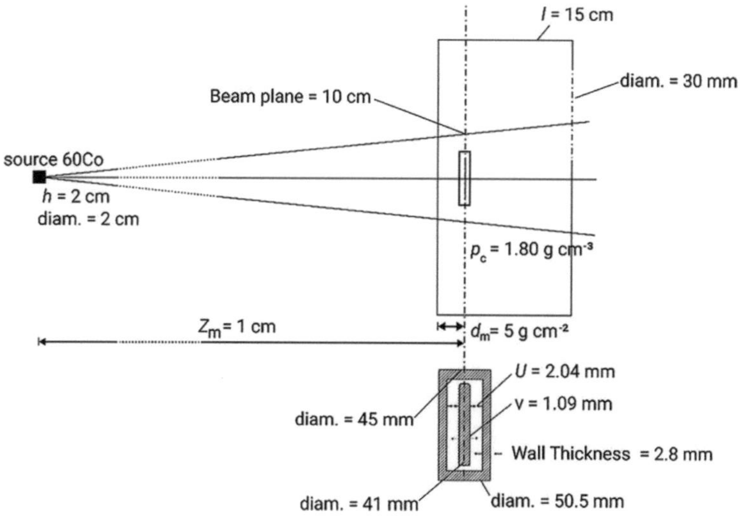

Figure 7.
Parallel plates graphite ionization chamber (1.8 gm/cm^3) with 2.8 mm wall thickness, inner diameter of 45 mm, outer diameter of 50.5 mm, used by the BIPM and reported by Boutillon and Niatel [18].

The reference conditions include radiation field of 10 x 10 cm^2 in the plane of the phantom surface, SSD = 100 cm, with the center of the chamber positioned at 5 g/cm^2 depth in graphite, reference air temperature of 22°C, atmospheric pressure of 101.3 kPa, and humidity between 30 and 70%, according to the formalism:

$$D_w = \frac{I}{\rho} \bullet v \bullet \frac{W_{air}}{e} \bullet \left(\frac{\mu_{en}}{\rho}\right)_{w,c} \bullet sc, a \bullet \Pi k_j \tag{6}$$

where:

I = current reading corrected for the reference conditions of T and P;

ρ = air density;

v = sensitive volume of the cavity;

$\frac{W_{air}}{e}$ = average energy needed to produce a pair of ions, its product being equal to the energy ceded to the air mass m_{air} from the reference sensitive volume;

$\left(\frac{\mu_{en}}{\rho}\right)_{w,c}$ = ratio between the mass-energy absorption coefficients for water and graphite. Proposed by Hubbel and Seltzer [15];

sc, a = ratio of the restricted stopping power between graphite and air, calculated based on the Spencer-Attix theory taking into account the average value of the energy in the electron spectrum generated by the effect;

Πk_j = the product of several correction factors:

kh = correction for the reference humidity;

ks = loss by ionic recombination;

km = radial non-uniformity of the beam in the chamber plane;

(d/do) = deviation correction between nominal and actual distance;

f = graphite to water conversion factor.

Fricke dosimetry consists of measuring the conversion, due to the ionizing radiation, of the ferrous ions present in the solution, into ferric ions through spectrophotometry. The Fricke dosimeter consists of a 96% water solution, therefore its attenuation to radiation is very similar to that of water and can be used in the dose range of 5 Gy–400 Gy with dose rates of up to 106 Gy/s.

The quantity determined by the Fricke chemical dosimetry system is the absorbed dose to the Fricke solution (D_F), as defined in Eq. (7) and described in the literature by [19, 20].

$$G\left(Fe^{3+}\right) = \frac{\Delta OD}{D_F.L.\rho.\varepsilon} \tag{7}$$

Where:

ΔOD = difference between the absorbance of the irradiated solution and the control solution, corrected for the temperature during irradiation and reading measured at 304 nm;

$G(Fe^{+3})$ = chemical yield of the reaction for the gamma radiation beam;

L = optical pathlength of the cuvette, where the solution is placed during the readings by the spectrophotometer;

ρ = density of the Fricke solution;

ε = molar absorptivity coefficient or molar extinction coefficient;

To determine the quantity of interest, D_w in water, it is necessary to use the correction factors defined in Eq. (8), as proposed by [21] and expanded by [19]:

$$D_w = D_F \bullet f_{w,F} \cdot P_{wall} \bullet f_{avg} \quad (8)$$

Where:

D_F = absorbed dose to the Fricke solution;

$f_{w,F}$= factor that converts the absorbed dose to the Fricke solution to the absorbed dose to water.

P_{wall}= factor that corrects disturbances caused by the PMMA walls of the holders containing the solution.

f_{avg} = factor that corrects the inhomogeneity of the dose deposited in the Fricke solution along the radial and the vertical axis.

This method requires laboratories with several parameters under control such as temperature, dust, cleaning, laminar flow hoods, Milli Q water production, glassware, quartz cuvettes, high-resolution double-beam spectrophotometer with filters for your QA, and high-purity chemicals. For this reason, its use is restricted to laboratories and not to be used at clinical environments.

4. Reference dosimetry

It refers to the measurement of the absorbed dose in water with an ionization chamber in the beam of the user's Institution. The reference conditions used in the calibration laboratory must reproduced, and the influence quantities (T, P, U) are measured at the time of data acquisition and correction accordingly.

Step 1: Calibration of a user's chamber at the level of the National Laboratory or of an SSDL according to interface [3].

$$ND_{w,Q} = \frac{{}^{lab}D_{w,Q}}{{}^{lab}M_{w,Q}} \quad (9)$$

where:

$ND_{w,Q}$ = calibration coefficient provided by SSDL or PSDL to the user;

${}^{lab}D_{w,Q}$ = absorbed dose to water determined in the SSDL by the standard instrument under reference conditions, that is, SSD = 100 cm, radiation field 10 × 10 cm^2 and the chamber centered at a depth of 5 cm in water;

${}^{lab}D_{w,Q}$ = reading of the user's chamber called reference chamber, performed on the same beam and under the same conditions as in the SSDL or PSDL.

Step 2. With the calibration coefficient $ND_{w,Q}$.

These measurements are performed at the user's institution with its reference chamber to obtain the absorbed dose to water with a beam of the same quality as the SSDL under the reference conditions: SSD =100 cm, radiation field 10 x 10 ^{cm2} and depth of 5 cm in water according to the Eq. (10):

$$ {}^{u}D_{w,Q} = {}^{u}M_{w,Q} \bullet N_{D,w,Q} \quad (10)$$

where:

${}^{u}D_{w,Q}$ = dose measured in the user's beam under reference conditions;

${}^{u}M_{w,Q}$ = average reading of the reference chamber in the user's beam;

$N_{D,w,Q}$ = calibration coefficient provided to the user for a given beam quality by the Calibration Laboratory, in general gamma rays of ^{60}Co.

As the calibration coefficient is normally defined for a ^{60}Co gamma ray beam, if the user has a different beam (e.g., photons with 6, 10, 15 MV) a *Kq* factor well described by Andreo et al. [6] should be used to adjust the detector's response to this new beam quality according to the Eq. (11):

$$^{u}D_{w,Q} = {}^{u}M_{w,Q} \bullet N_{D,w,Q} \bullet k_Q \tag{11}$$

where:

$^{u}D_{w,Q}$ = dose measured in the user's beam under reference conditions;

$^{u}M_{w,Q}$ = average reading of the reference chamber in the user's beam;

$N_{D,w,Q}$ = calibration coefficient provided to the user for a given beam quality by the Calibration Laboratory, in general gamma rays of ^{60}Co.

k_Q = factor that adjusts the value measured in the quality of the user's beam defined from the relationship between the readings taken on the water phantom, with a 10 x 10 cm^2 radiation field size defined at 20 cm and measured at 10 cm in depth in the same geometry, that is, according to the definition of the $TPR_{20,10}$ as shown in **Figure 8**.

The numerical value of this factor varies with the type of materials used in the chambers, whose beam quality is expressed by the $TPR_{20,10}$ ratio, which empirically represents the variation in the interaction and absorption behavior of each of the materials due to the different cross sections. Typical behavior of *Kq* values as a function of the beam quality, defined by the $TPR_{20,10}$, is shown in **Figure 9**.

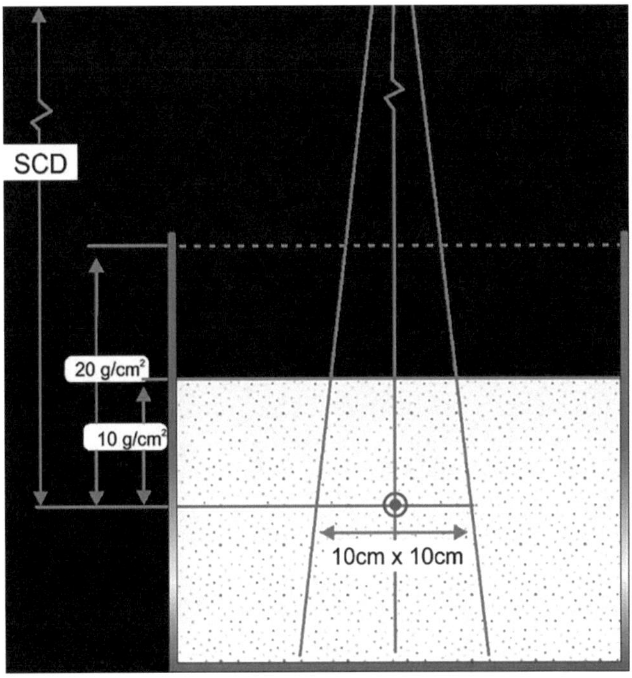

Figure 8.
Geometry that should be used for measurement of the quality of the Q beam, to obtain the k_Q factor from the $TPR_{20,10}$ ratio, for a source chamber distance (SCD) of 100 cm, 10 x 10 cm^2 field and measurements at depths of 10 and 20 g/cm^2 of water as recommended by the TRS#398 [1].

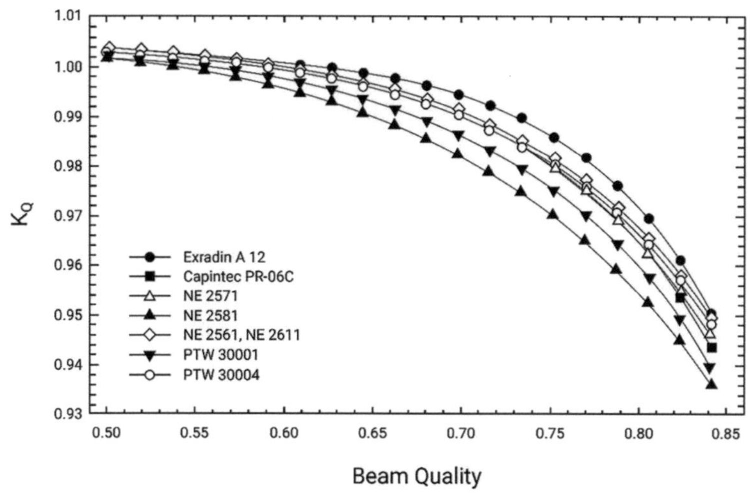

Figure 9.
Typical behavior of Kq *values as a function of the beam quality, defined by the* $TPR_{20,10}$.

The graph clearly shows a dependence of the *Kq* value with the type of the chamber, in this case for photons of different energies, using Farmer-type cylindrical chambers from various manufacturers, built with different materials. TRS#398 [1].

The measurement system that best suits this application at the user level is the ionization chamber, in which case there is no need to know its volume as the calibration coefficient considers the chamber's response and not its real volume.

The $TPR_{20,10}$ can also be estimated from the Percentage Depth Dose measurements using the empirical relationship, according to Eq. (12):

$$TPR_{20,10} = 1.2661 \bullet PDD_{20.10} - 0.0595 \tag{12}$$

where,

$TPR_{20,10}$= ratio of ionization measurements at 20 cm and 10 cm depth in water for a constant source to chamber distance and with a 10 x 10 cm field at the plane of the detector.

$PDD_{20.10}$ = ratio between the values measured at 20 and 10 cm depth for a 10 x 10 cm^2 field at a source camera distance of 100 cm.

5. Relative dosimetry

In the clinical environment various measurements are performed under non-reference conditions where the calibration coefficient does not need to be used. These measurements are called relative, such as: dosimetry of other radiation fields (values compared with the reference field, output factors), wedge filter factor (ratio between readings performed with and without filter on the same geometry), measurements of depth dose (normalized to the values obtained at the maximum dose point for that specific radiation field and type of beam).

In these cases, there is a variety of detectors that can be used without compromising on having their values related to the true value of the quantity.

For example: diodes, TLDs, micro-cameras, detector array, alanine, film, MOSFET among others, all of them with their well-defined and different characteristics, such as (sensitivity, short term repeatability, long-term stability, angular, dose rate and energy dependence, detector size, leakage, signal fading) among others must be considered.

6. Key points

Check and consider, if applicable, the following:

- energy dependence with depth of water.
- Dose rate dependence, especially on FFF (flattening filter-free) beams.
- Directional dependence due to the detector geometry and volume.
- Signal-to-noise ratio as a function of field size, detector shape and size, and signal sensitivity.
- Permanent defects caused by dose storage
- volume that results in loss of spatial resolution.

Special cases where the reference conditions are not able to follow TRS#398 [1] recommendations are called non-reference conditions. Small fields used in radiosurgery show a more complex spectrum and require ionization chambers with other dimensions, additional geometric conditions, and specific formalism.

In this case, the TRS# 483 [20] should be used as a reference, the most suitable one at this time, where a relatively small variety of detectors are used, generally limited by the field size and the loss of lateral electronic balance.

Replace the entirety of this text with the main body of your chapter. The body is where the author explains experiments, presents, and interprets data of one's research. Authors are free to decide how the main body will be structured. However, you are required to have at least one heading. Please ensure that either British or American English is used consistently in your chapter.

7. Conclusion

This entire chain of measurements and formalism must take into account the specific physical conditions of the interaction processes between the radiation beam with the detector in the measurement processes, aiming to ensure the least possible uncertainty in the dose delivered to the patient.

The different levels of complexity and duties of the metrological stakeholders are a result of the complexity of the experimental arrangements, the quality of the measurement systems, the degree of control over the environmental conditions and the high cost, which makes it not compatible with the clinical environment.

However, the metrological consistency between the different levels guarantees a level of final uncertainty of the dose delivered to the patient compatible with the recommendations of international organizations.

Therefore, if we keep the instruments (electrometer + cable + camera) accompanied by a quality assurance program, with its periodic calibrations and care to maintain its functional integrity, the final quality of the measurements will always be in accordance with the concept of the best practice.

Conflict of interest

The authors declare no conflict of interest.

Author details

Carlos Eduardo de Almeida[1] and Camila Salata[2*]

1 Radiological Sciences Department, Rio de Janeiro State University (UERJ), Rio de Janeiro, Brazil

2 Department of Medical and Research Facilities, National Nuclear Energy Authority (CNEN), Rio de Janeiro, Brazil

*Address all correspondence to: salata@cnen.gov.br

References

[1] IAEA. Technical Report Series TRS#398, Absorbed dose determination in external beam radiotherapy. Vienna, Austria: IAEA; 2010

[2] IAEA. Technical Reports Series 469-Calibration of Reference Dosimeters for External Beam Radiotherapy. Vienna, Austria: IAEA; 2009

[3] Andreo P, Burns DT, Nahum AE, Seuntjens J, Attix FH. Fundamentals of Ionizing Radiation Dosimetry. 1st ed. Weinheim, Germany: Wiley-VCH; 2017

[4] Radiation quantities and units ICRU report 33. United States: International Commission on Radiation Units and Measurements; 1980

[5] Peixoto JGP, de Almeida CE. Implementation of a Brazilian primary standard for x-ray. In: International Symposium on Standards and Codes of Practice in Medical Radiation Dosimetry. Vienna, Austria: NIST; 25–28 November 2002

[6] Allisy, Contribution à la mesure de l'exposition produite par les photons émis par le Cobalt 60. Metrologia: IAEA; vol. 3, no. 2, pp. 41–51, 1967. DOI: 10.1088/0026-1394/3/2/003.

[7] Boutillon M, Niatel M-T. A study of a graphite cavity chamber for absolute measurements of 60Co gamma rays. Metrologia. 1973;**139**:146. DOI: 10.1088/0026-1394/9/4/001

[8] Buckley LA, Rogers DW. Wall correction factors, Pwall, for thimble ionization chambers. Med Phys. 2006; **33**(2):455-464. DOI: 10.1118/1.2161403

[9] de Almeida CE, Niatel MT. Comparison Between IRD and. BIPM Exposure and Air-Kerma Standards for Cobalt-60 Gamma-. Rays. Sèvres: Bureau International des Poids et Mesures (BIPM); 1986. pp. 86-92. Rapport BIPM-1986/12

[10] de Almeida CE, Malamut C, Rodrigues LN. Experimental arrangement and data acquisition system at the LNMRI for exposure and air kerma measurement of Cobalt-60 gamma rays. Journal of Medical Physics. 1996;**21**:1-5

[11] de Almeida CE, Rodrigues LN, Cecatti ER, Malamut C. Exposure and air-kerma standards for cobalt-60 gamma rays. Revista de Física Médica Aplicada e Instrumentação. 1990;**5**:211-228

[12] Duftschmid K. Private Communication. 1986

[13] Hubbell JH, Seltzer SM. Tables of x-ray mass attenuation coefficients and mass energy-absorption coefficients 1 keV to 20 meV for elements z = 1 to 92 and 48 additional substances of dosimetric interest. United States: IAEA; 1995

[14] Allisy-Roberts PJ, Boutillon M, Rodrigues LN. Comparison of the standards of air kerma of the LNMRI and the BIPM for 60Co γ-rays. Bureau International des Poids et Mesures. 1996, Rapport BIPM-96/3.

[15] Alissy-Roberts PJ, Kessler C, Mello da Silva CN. Comparison of the standards of air-kerma of the LNMRI and the BIPM for 60Co gamma rays. Bureau International des Poids et Mesures. 2005, Rapport BIPM-2005/01

[16] Malcolm E. Primary standards of air kerma and absorbed dose AAPM. Summer School. Medical Physics Publisher; 2009

[17] Rosado PH, De Almeida CE. Water Calorimeter for absolute determination of absorbed dose in water. Work in Progress

[18] Boutillon M, Niatel MT. A study of graphite chambre for absolute measurements of 60Co gamma Rays. Metrologia. 1973;**9**:139-146. DOI: 10.1088/0026-1394/9/4/001

[19] De Almeida CE, Ochoa R, de Lima MC, David MG, Pires EJ, Peixoto JG, et al. A feasibility study of fricke dosimetry as an absorbed dose to water standard for 192Ir HDR sources. PLoS One. 2014;**9**:e115155. DOI: 10.1371/journal.pone.0115155

[20] Alfonso R et al. A new formalism for reference dosimetry of small and nonstandard fields. Med Phys. 2008;**35**: 5179–5186

[21] Klassen NV, Shortt KR, Seuntjens J, Ross CK. Fricke dosimetry: The difference between G(Fe3+) for 60Co γ-rays and high-energy x-rays. Phys Med Biol. 1999;**44**:1609-1624. DOI: 10.1088/0031-9155/44/7/303

Chapter 5

Thermoluminescence Dosimetry Technique for Radiation Detection Applications

Hossam Donya

Abstract

Due to the risk of radiation exposure, radiation dosimetry is performed regularly to ensure the occupational safety of personnel and radiation workers. Therefore, various dosimeters are widely used to detect neutrons, gamma, X-ray, and proton irradiation fields. As an example, in medical applications, routine personal dosimetry is used to monitor and limit workers' long-term occupational exposure. Radiation workers who undertake X-ray diagnostic, radiotherapy operations, in clinical and industrial application. Although, the overheads of running an in-house TLD (Thermoluminescent dosimetry) service for monitoring doses to eyes, pacemakers and so on seems rather high for the benefits conferred, however, it is still widely used for reporting doses accurately in various medical centers over the world. TLD also is widely used for measuring entrance doses on a handful of patients to validate a new LINAC/TPS combination. As well as in the industrial field as if petroleum, companies or nuclear reactor, RSO (radiation safety officer) used TLD badges to report delivered doses. In this chapter, we focus on the TLD technique for measuring doses of various ionizing radiation detection. Different methods for evaluations of TL Kinetics are covered. Modern TLD applications in the clinical field are also investigated. Some recommendations on advance dosimetry failure of TLD are concluded.

Keywords: TLD, TL kinetics, radiotherapy, hybrid-functional density theory, modern clinical applications

1. Introduction

TLD method is considered an important technique as it can store radiation in trap centers for long period. Glow peaks of thermoluminescence dosimeters are later measured and discussed based on some models related to the physical changes in the band structure of dosimeter because of ionizing radiation exposure. A wide range of substances exhibits thermoluminescence (TL) phenomena after being exposed to nuclear radiation such as activated LiF and CaSO4. Thermoluminescent dosimeter (TLD) emits light when heated up after being irradiated. Due to this special property, TLD could be used as a radiation dosimeter. TLD has many advantages and sensitive to different types of radiation. A dosimeter of higher TL response to thermal neutrons is most commonly used in mixed radiation fields (neutron and gamma ray). The sensitivity of TLD to neutrons depends on the detector compound type, environment and neutron energy. For neutron dosimetry purposes, the neutron and gamma ray dose contribution must be separated by using two different detector types of TLD. The first one should be sensitive to gamma and the other should be sensitive to neutrons plus gamma (as LiF-700 and LiF-600) [1, 2].

The response of fast neutrons depends on the cross-section for the interaction in TLD material and the relative TL efficiency, which depends on the linear energy transfer (LET) of the reaction products in the first place. The response to intermediate-energy depends mainly on the cross-section of the reaction, which may take place with the composite material of the TLD.

1.1 TLD applications in neutron and gamma ray dosimetry

Generally, there are three types of TLD used for neutron dosimetry as follow:

1.1.1 Albedo neutron dosimeter

A considerable fraction of intermediate and fast neutrons can be slowed down to epithermal neutron energy and backscattered in the human body, interacting with the sensitive TL material. An albedo neutron dosimeter is a type of neutron monitor and is typically used in the neutron energy range of 0.2 eV to around 0.5 MeV. The slow neutrons interact with TL material, usually through ^{6}Li (n,α) 3H reaction, and the resulting induced charged particles to stimulate the TL material. Recently, some of albedo TLD dosimeters depend on ^{10}B (n, α) 7Li reactions. Because neutron TL sensitive material responds to gamma radiation, and neutrons are accompanied by this gamma radiation, another TLD is usually utilized in conjunction with TLD with a gamma ray.

The neutron albedo dosimeter measures (a) direct fast neutrons, (b) direct thermal neutrons, and (c) albedo neutrons reflected from the body. This type of dosimeter uses Lexan polycarbonate and/or CR-39 foils, as well as two 10B (n, γ) ^{7}Li converters in a cadmium cover, to efficiently measure the three neutron dosage components independently [3–5]. Fast neutron dose is assessed in CR-39 by counting proton recoil tracks, while thermal neutron dose is determined by counting α particles created during the process. Because the albedo dosimeter has a sensitivity range of 0.3–30 mSv, it is advised that it be used as a backup dosimeter to assist in the assessment of high dose values in the event of accidents or patients receiving neutron therapy.

In another application, the ^{10}B (n, α) ^{7}Li reactions with the backscattered albedo neutrons employed with Electret's ionization chamber proposed by Seifert et al. [6, 7]. In this chamber, induced ^{7}Li from the ionization of the gas in the chamber worn on the body's surface in the above reaction instance. Under saturation conditions, produced charge carriers with the corresponding polarity travel to the surface of the electret. As a result, the change in the electrets voltage is a direct measure of albedo neutron fluence and an indirect estimate of primary neutron fluence. In general, the advantages of albedo TLD dosimeter are: they are relatively inexpensive and can be reused, easily fabricated, lightweight to wear, Readout is simple and can be automated, Insensitive to humidity.

While their disadvantages are: Some of TLD exhibit fading, TLD is sensitive to gamma-ray, they must be worm properly or serious errors can be resulted, the measured values of TLD does not give permanent record as the track detectors, their sensitivity is highly dependent on the angle and energy of incidence radiation.

1.1.2 Hydrogenous radiator TLDs

In this type of dosimeters, the fast neutrons knock out protons from hydrogenous material mixed with the phosphor, and the protons dispel their energy in the dosimeter. In this method, the hydrogenous substances are called proton radiators [8]. This technique has demonstrated that TL materials mixed with hydrogenous material can detect fast neutrons, but the sensitivity needs to be improved by one order of magnitude before using in personnel neutron dosimetry.

1.1.3 LET-dependent deep trap TLD glow peaks

The fast neutron interacts directly with the TL material as calcium fluoride (CaF_2: Tm) which is commercially called TLD-300. This type has a glow curve with two glow peaks and the peak temperature T_m centered 150 and 250°C, respectively. The higher temperature peak (250°C) has a greater response to the fast neutrons. TLD-300 dosimeter CaF_2: Tm (0.35 Mol. %) showed a lower detection limit of about 0.3 mSv from ^{241}Am-Be source.

2. Characteristic of TLD phosphors

2.1 The glow curve

The term "Glow curve" refers to the graph of TL as a function either of temperature or of time as shown in **Figure 1**.

Glow curves have the following features:-

- The glow curve of a certain phosphor probably best characterizes that phosphor. For example, the appearance of glow peaks only at low temperatures implies that the phosphor loses its stored TL with time, and therefore would be unsuitable for long-term measurements.

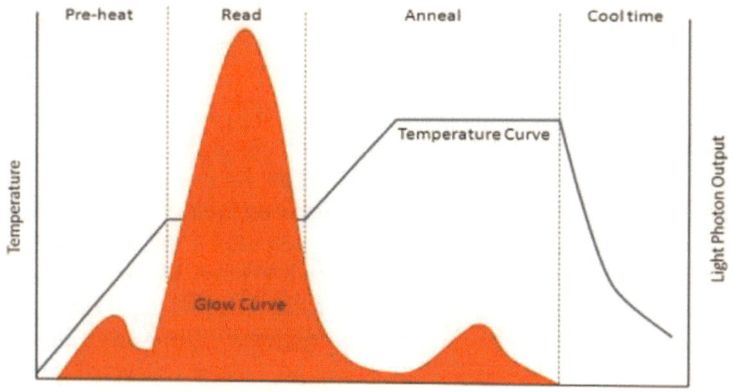

Figure 1.
TLD glow curve and time–temperature profile (TTP).

- A peak at very high temperatures indicates a phosphor that will produce infrared radiation at the temperature necessary to release the TL. This is a problem for the instrumentation.

- A glow curve without well-defined peaks makes the selection of the appropriate end for integration difficult. Ideally then, a glow curve should reveal only a single thermoluminescent peak, which occurs at a temperature high enough to ensure room temperature stability but not so high as to present instrumental problems. However, the temperature at which a TL peak appears is quite affected by the heating rate.

The following factors may affect the shape of the glow curve:

2.2 TL sensitivity

The sensitivity of TLD should be evaluated for competitor's materials to determine the dose linearity, sub linearity or supralinearity behavior of phosphors. The sensitivity and thermal stability of glass samples were found to be dependent on both the starting materials and the method of preparation in Refs. [9–13].

2.3 Dose rate dependence

TL dosimeters, in general, have demonstrated no dose rate effects over a wide range.

2.4 Stability

TL is the release, in the form of visible light, of energy absorbed from previous irradiation. The rate at which this energy is released is dependent upon the phosphor temperature and increases sharply at a higher temperature. Even though the concept of "glow peak temperature" that temperature at which the maximum of the glow peak occurs, is both useful and easily demonstrated. It should be remembered that a finite rate of loss of stored energy exists even for much lower temperatures. If a phosphor shows an insignificant loss of TL at room temperature, it is said to have good stability.

3. TL kinetics

The physical process leading to the emission of TL from a sample is related in most cases to the traffic of charge carriers, usually electrons and holes, between different imperfection states in the solid sample. Studying the kinetics of the TL process means the investigation of electron–hole transitions between energy states in cases of both the irradiation of the TL sample and the readout processes. Although, in most experimental situations the TL curve consists of several overlapping peaks, it is appropriate to start the discussion by dealing with a single peak to understand the basic process. For most purposes, it is not necessary to assume that the glow curve consists of only one peak. The analysis of a single peak may just be valid if a series of peaks occur, provided that the peak of interest is sufficiently separated from others, either because it appears separately or because we have an efficient method to isolate it from the rest of the curve.

3.1 First kinetics order

If n is the number of trapped electrons in the sample, which is maintained at constant temperature T, n decreases with time t as:

$$\frac{-dn}{dt} = \mathrm{P\,n} \tag{1}$$

From Eq. (1) we get:

$$\frac{-dn}{dt} = nS\exp\left(\frac{-E}{KT}\right) \tag{2}$$

The rate of photon emission, and hence the rate of release of electrons from traps to their rate of arrival at luminescence centers, determine the strength of the TL glow peak [14].

$$I\alpha - \frac{dn}{dt} \to I = -C\frac{dn}{dt} = nCS\exp\left(\frac{-E}{KT}\right) \tag{3}$$

Where C is a luminescence efficiency constant.

When the dosimeter is heated with rate $\beta = dT/dt$. Then we may write dn/dt as:

$$\frac{dn}{dt} = \frac{dn}{dT}\frac{dT}{dt} = \beta\frac{dn}{dT} \tag{4}$$

Substitute in Eq. (2), we get:

$$\to \frac{dn}{dT} = \frac{-nS}{\beta}\exp\left(\frac{-E}{KT.}\right)$$

$$\to \int_{n_o}^{n}\frac{dn}{n} = \frac{-S}{\beta}\int_{T_o}^{T}\exp\left(\frac{-E}{KT}\right)dT$$

Where n_o denotes the number of electrons in the trap at a given time and temperature, t_o and T_o, respectively.

$$n = n_o\exp\left(\frac{-S}{R}\int_{T_o}^{T}\exp\left(\frac{-E}{KT}\right)dT\right) \tag{5}$$

Substitute in Eq. (3)

$$I = n_oSC\exp\left(\frac{-E}{KT}\right)\exp\left(\frac{-S}{\beta}\int_{T_o}^{T}\exp\left(\frac{-E}{KT}\right)dT\right) \tag{6}$$

It should be noted that $I(T)$ in Eq. (6) dependents on two physical parameters, the activation energy E, and frequency factor S, and the heating rate β. The activation energy is the minimum energy required to release the electrons from their traps. Differentiation of Eq. (6) with respect to the temperature gives:

$$\frac{\partial I}{\partial T} = n_oSC\left\{\exp\left(\frac{-E}{KT}\right)\exp\left(f(T)\right)\frac{\partial f(T)}{\partial T} + \exp\left(f(T)\right)\exp\left(\frac{-E}{KT}\right)\left(\frac{E}{KT^2}\right)\right\}$$

Where, $f(T) = \frac{-S}{\beta}\int^{T} \exp\left(\frac{-E}{KT}\right) dT$

At $T = T_m \rightarrow \frac{\partial I}{\partial T} = o$

$$\rightarrow \frac{s}{\beta} \exp\left(\frac{-E}{KT_m}\right) = \frac{E}{KT_m^2} \tag{7}$$

where T_m is peak position or the temperature at maximum intensity.

Equation (7) describes the condition of the occurrence of the maximum intensity and the determination of the corresponding temperature, which we call, T_m. The reduction in the second exponential function is faster than the growth in the first exponential function above this temperature, and the product function decreases until the traps are fully depopulated. This accounts for the end of the peak. A theoretical (calculated) glow peak plotted using Eq. (6) is shown in **Figure 2**. The main feature of the first-order peak is that the asymmetric, is such that at temperatures over T_m, the reduction is faster than the rise at low temperatures.

The initial concentration n_o appears in the first kinetics order acts only as a constant multiplying the temperature-dependent factors. In this particular case of the first kinetics order, changing the initial concentration n_o has no effect on the curve's form because adjusting the intensity at each temperature has the same proportional effect. **Figure 3** shows several glow peaks with different n_o. One of the aspects of this fact is that T_m is independent of the initial concentration n_o.

This appears well in the condition of the T_m described by Eq. (7), where n_o does not appear in the equation. This property of the independent of T_m on n_o is specified to the first-order case, and will not occur for most of the other kinetics possibilities [13]. Eq. (7) can be written in the following form:

$$\frac{\beta E}{K} = ST_m^2 \exp\left(\frac{-E}{KT_m}\right) \tag{8}$$

We see that changing the heating rate β must change T_m in a such way that equality still holds. The term $T_m^2 \exp\left(\frac{-E}{KT}\right)$ is monotonically increasing with T_m, therefore increasing the rate β will immediately cause T_m to increase. Since $T_m^2 \exp\left(\frac{-E}{KT}\right)$ is a very rapidly increasing function of T_m, only a small change of T_m may accompany a large variation in the heating rate β, this variation is usually rather easily observable.

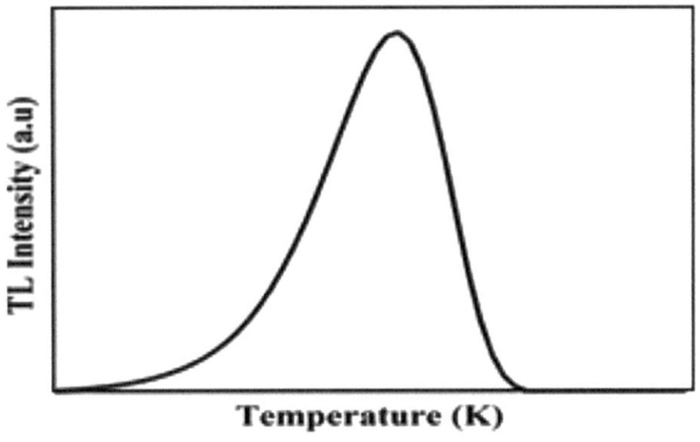

Figure 2.
Theoretical glow peak plotted using the first kinetics order equation.

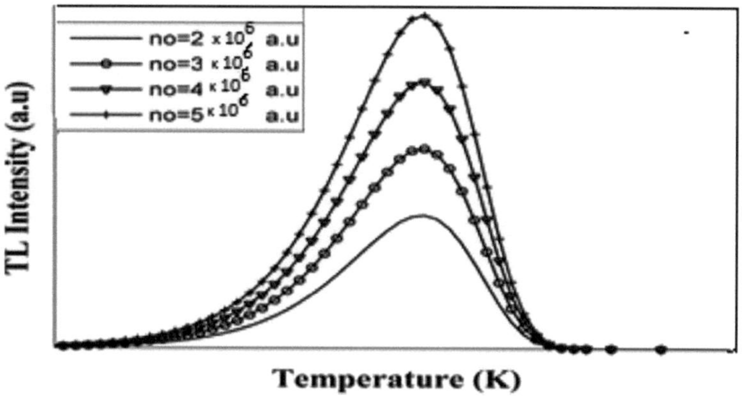

Figure 3.
Glow curves plotted using the first-order kinetics equation for different n_o.

3.1.1 Second kinetics order

One assumption made up by Randall and Wilkins [14] which led to the first kinetics order was that once a charge carrier is thermally elevated into the band, it is bound to recombine rather quickly with an opposite sign carrier trapped in a recombination center. Gralick and Gibson [15] considered another case in which the free carriers may re-trap with equal retrapping recombination probabilities with the further assumption that the concentration of electrons in traps and holes in recombination centers are equal during the entire process. Denoting the total number of traps of the given type (free electrons or holes) by N, they found the kinetics equation:

$$I = \frac{-\mathrm{d}n}{\mathrm{d}t} = \left(\frac{S}{N}\right)n^2 \exp\left(\frac{-E}{KT}\right) \tag{9}$$

where (S/N) is a constant having units of $\mathrm{m}^3\mathrm{s}^{-1}$, which we may denote by S'. Then we have

$$I = \frac{-\mathrm{d}n}{\mathrm{d}t} = S'n^2 \exp\left(\frac{-E}{KT}\right) \tag{10}$$

where S' is called "pre-exponential factor" which does not have the same meaning of "frequency factor" as was in the first kinetics order.

For linear heating rate β, we have:

$$I = \frac{-\mathrm{d}n}{\mathrm{d}t} = \frac{-\mathrm{d}n}{\mathrm{d}T} \underset{\rightarrow}{\text{yields}} \frac{\mathrm{d}n}{\mathrm{d}T} = \frac{-\acute{S}}{\beta} n^2 \exp\left(\frac{-E}{KT}\right)$$

$$\underset{\rightarrow}{\text{yields}} \frac{1}{n} - \frac{1}{no} = \frac{\acute{S}}{-}\int^{T} \exp\left(\frac{-E}{KT}\right)\mathrm{d}T$$

$$\underset{\rightarrow}{\text{yieldsn}} = n_o\left\{1 + \left(\frac{\acute{S}}{-}\right)n_o\int^{T} \exp\left(\frac{-E}{KT}\right)\mathrm{d}T\right\}^{-1} \tag{11}$$

Substitute in Eq. (10),

$$I = \acute{S}n_o^2 \exp\left(\frac{-E}{KT}\right)\left\{1 + \left(\frac{\acute{S}}{-}\right) n_o \int^T \exp\left(\frac{-E}{KT}\right) dT\right\}^{-2} \tag{12}$$

where Eq. (12) represents the intensity of a glow peak according to the second kinetics order model. At high temperature, the second decreasing function dominates so that the product function is decreasing. Somewhere between two regions the glow curve, therefore, reaches its maximum. **Figure 4** Displays a hypothetical glow peak plotted using Eq. (12).

The condition of the maximum is found by setting the derivative of Eq. (12) to zero (dI/dT = zero) [16], then we may find:

$$\frac{dI}{dT} = \acute{S}n_o^2 \left\{ \begin{array}{l} -2\,exp\left(\frac{-E}{KT}\right)\left\{1 + \left(\frac{\acute{S}n_o}{\beta}\right)\int^T \exp\left(\frac{-E}{KT}\right)dT\right\}^{-3}\left(\left(\frac{\acute{S}n_o}{\beta}\right)\exp\left(\frac{-E}{KT}\right)\right) \\ +E/KT^2 \exp\left(\frac{-E}{KT}\right)\left\{1 + \left(\frac{\acute{S}n_o}{\beta}\right)\int^T \exp\left(\frac{-E}{KT}\right)dT\right\}^{-2} \end{array} \right\}$$

$$\frac{dI}{dT} = 0 at T = T_m \underset{\rightarrow}{yields} \frac{2\acute{S}n_o}{\beta} \exp\left(\frac{-2E}{KT_m}\right)\left\{1 + \left(\frac{\acute{S}n_o}{\beta}\right)\int^{T_m} \exp\left(\frac{-E}{KT_m}\right)dT\right\}^{-3}$$

$$\frac{E}{KT_m^2} \exp\left(\frac{-E}{KT_m}\right)\left\{1 + \left(\frac{\acute{S}n_o}{\beta}\right)\int^{T_m} \exp\left(\frac{-E}{KT_m}\right)dT\right\}^{-2}$$

Multiply by $\left\{1 + \left(\frac{\acute{S}}{\beta}\right) no \int^{T_m} \exp\left(\frac{-E}{KT_m}\right)dT\right\}^3$ and rearrange, one gets

$$\left\{1 + \left(\frac{\acute{S}}{\beta}\right) no \int^{T_m} \exp\left(\frac{-E}{KT_m}\right)dT\right\} = \frac{2KT_m^2 \acute{S}n_o}{\beta E} \exp\left(\frac{-E}{KT_m}\right) \tag{13}$$

Then Eq. (13) represents the condition of the peak maximum according to the second kinetics order. As can see n_o appears in the equation and therefore we expect that T_m will depend on n_o. It can be shown numerically or analytically, that increasing

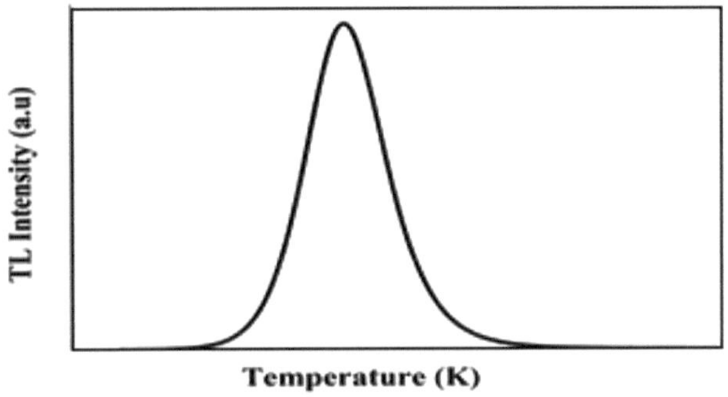

Figure 4.
Theoretical glow peak plotted using the second-order kinetics equation.

n_o causes T_m to decrease. An exception to this rule of the shift of a second-order peak with n_o can be found by Wrzesinska [17], who writes Eq. (10) with $S' = \frac{S}{n_o}$. The resulting peak has all the regular features of a second-order peak (e.g., symmetry properties) except one can write S instead of n_oS' and thus Eq. (10) turns out to be independent of n_o. The ensuring T_m is, therefore independent of n_o. It is not clear, however, what physical circumstances result in S' being equal to S/n_o [17]. Other aspects of the dependence of the glow curve on the initial concentration n_o are paramount importance when we are interested in a TL as a dosimetric tool. In many cases, one associated the initial concentration with the imparted dose and then the dependence of different parts of the glow peak on n_o is important. In the first kinetics order, since the intensity at each point is multiplied by the same factor while changing n_o, the total area varies with the same amount so that the total area is proportional to n_o. Its occurrence in second-order peak can be illustrated by integrating Eq. (9) with respect to time from zero to infinity;

$$\int_0^\infty I(t)\mathrm{d}t = -\int_{n_o}^{n} \mathrm{d}n = n_o - n_\infty = n_o \tag{14}$$

Both in the first order and second order, as well as other cases, n_∞ is zero and therefore the integral, which represents the area under the glow peak is equal (in appropriate units) to n_o.

Now we can consider the dependence of different portions of the second-order peak on n_o. First, we shall study the dependence of I on n_o for a given temperature T. In the initial rise range, Eq. (12) reduces to:

$$I(T) \cong n_o^2 S' \exp\left(\frac{-E}{KT}\right) \tag{15}$$

This shows immediately that for a given temperature in this range the dependence of I on n_o is superlinear, namely $I \alpha n_o$. It is to be emphasized that it is true only in the initial rise region; as already shown the total area is proportional to n_o and different dependencies are expected on other portions of the curve. Using the maximum condition equation and approximation to $\int^T \exp(-E/KT)\mathrm{d}T$, it can be shown that the two terms in the brackets in Eq. (12), namely unity and $n_o \frac{S'}{\beta} T_o \exp(-E/KT)\mathrm{d}T$ are more or less equal at $T = T_\mathrm{m}$. At higher temperature, the latter term increases substantially and the unity can be neglected so that we obtain:

$$I \cong S' no^2 \exp\left(\frac{-E}{KT}\right)\left\{\left(\frac{S'}{\beta}\right)\mathrm{no}\int^T exp\left(\frac{-E}{KT}\right)\mathrm{d}T\right\}^{-2}$$

$$\underset{\rightarrow}{\mathrm{yieldsI}} \cong S' \exp\left(\frac{-E}{KT}\right)\left\{\left(\frac{S'}{\beta}\right)\int^T \exp\left(\frac{-E}{KT}\right)dT\right\}^{-2} \tag{16}$$

The main point in Eq. (18) is that the term includes n_o cancel. This means that at a higher temperature range the TL intensity is independent of n_o for any given temperature [17].

Figure 5 shows plotted glow peaks using Eq. (12) for different n_o. In the low-temperature range, the TL intensity appears to depend on n_o. As n_o increases, T_m decreases which makes the peaks appear to be shifted to the low-temperature side. As the temperature increases the effect of n_o on the peak starts to decrease which makes the peaks approach each other's on the high-temperature side.

3.1.2 A single TL peak analysis

As seen in **Figure 6**, the concentration of the trapping state is denoted by N (m^{-3}), with $n(t)$ (m^{-3}) being filled by electrons at time t(s). These electrons can be thermally elevated into the conduction band by crossing an energy barrier of E (eV) at a rate proportional to exp.($-E$/kT), resulting in a concentration of free electrons $n_c(t)$ (m^{-3}). Following that, these can be retrapped in a similar trap with a re-trapping probability A_n or recombined with a trapped hole in a center with a recombination center probability A_m, generating a photon with the recombination center energy h. A set of three simultaneous differential equations governs this operation. The following factors influence the recombination process:

$$\underset{\rightarrow}{\mathrm{I\alpha n_c m yields}} I = \frac{-dm}{dt} = A_m n_c m \tag{17}$$

where n, m, and n_c are the trapped electron, hole in the center, and free-electron concentrations, respectively, and (dm/dt) is the recombination rate. This means that the amount of light emitted is proportional to the pace at which m decreases. The rate of recombination is proportional to both the instantaneous concentration of free

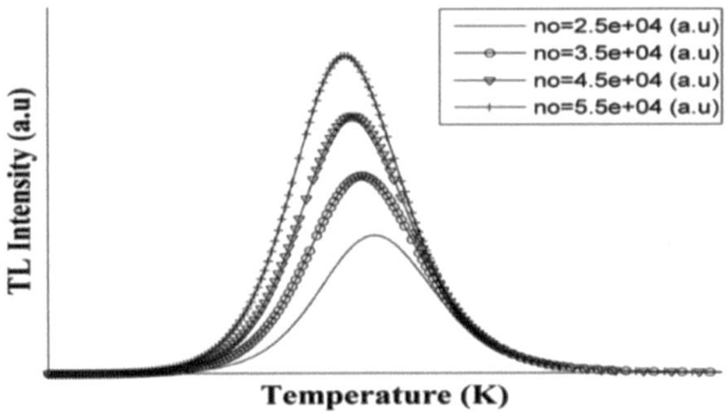

Figure 5.
Plotted glow peaks using the second-order kinetics equation for different n_o.

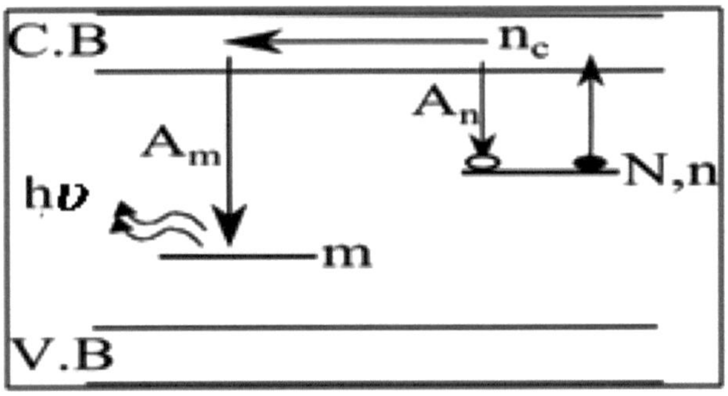

Figure 6.
A general treatment of the charge carriers' transitions in the TL sample.

electrons n_c and the concentration of hole centers m, the proportional constant Am ($m^3\ s^{-1}$). The product of cross-section recombination σ (m^2) and thermal velocity is commonly used to calculate this value ($m.s^{-1}$). The second equation is concerned with the movement of electrons that have been thermally liberated from the trapped condition. The rate of release of these electrons $-dn/dt$ is proportional to the trapped electron concentration n (m^{-3}) and the Boltzman constant exp.($-E/KT$), with S serving as the proportional constant (s^{-1}).

However, the actual rate of change of n is also related to the retraping term. The rate of retraping is proportional to the concentration of free electrons n_c, and the unoccupied trapping states N-n, the proportional factor being the recombination probability $A_n(m^3s^{-1})$. Thus, the second combined equation is given by:

$$\frac{-dn}{dt} = Sn \exp\left(\frac{-E}{KT}\right) - A_n n_c (N-n) \tag{18}$$

The third equation is that of charge neutrality. In its simplest form, it should read $m = n + n_c$. Taking the first derivative with respect to time, the charge neutrality condition can be written as:

$$\frac{dm}{dt} = \frac{dn}{dt} + \frac{dn_c}{dt} \tag{19}$$

$$\underset{\rightarrow}{\text{yields}} \frac{dn_c}{dt} = Sn \exp\left(\frac{-E}{KT}\right) - n_c\{mA_m + (N-n)A_n\} \tag{20}$$

This equation has been given by Adirovitch [18] for phosphorescence and by Halperin and Braner [19].

Now let us discuss the kinetics of the process in more general terms and see how the simplified cases of first, second, and more general cases emerge from Eqs. (17)–(20). Two simplifying assumptions were first made by Adirovitch [18] and later by many other investigators [19–23]. These are related to the relation between the concentration of the electrons in the conduction band and in traps and to the rate of change of these concentrations, namely:

$$\left|\frac{dd}{dt}\right| \ll \left|\frac{dn}{dt}\right|, n_c \ll n \tag{21}$$

Although, it seems to be the same connection between these two conditions, basically they are two separate relations and the occurrence of one does not necessarily imply the other. With these assumptions, Helperin and Braner [19] found the expression:

$$I = \frac{-dm}{dt} = \frac{mA_m}{mA_m + A_n(N-n)} Sn \exp\left(\frac{-E}{KT}\right) \tag{22}$$

Since this equation contains two unknown functions, $n(t)$ and $m(t)$, it cannot be solved without further assumption. As mentioned, Randall and Wilkins [14] wrote their first-order equation assuming strong recombination. This can be expressed in more specific terms. If we assume with relation to Eq. (22) that:

$$mA_m \gg (N-n)A_n$$

The condition of Eq. (22) is the relation between functions rather than parameters. It is, therefore, possible that at the low-temperature range of a glow peak, the strong inequality holds, and at higher temperatures where m and n decreases, the inequality “weakens” may be inverted. This may result in a shift from first-order behavior to non-first-order behavior within the same peak [24].

Then, one can say that:

$$mA_m + (N-n)A_n \cong mA_m \underset{\rightarrow}{\text{yields}} \frac{mA_m}{mA_m + A_n(N-n)} \cong 1$$

Then Eq. (22) will take the following form:

$$I = \frac{-\mathrm{d}m}{\mathrm{d}t} = Sn\exp\left(\frac{-E}{KT}\right) \tag{23}$$

Then we see that Eq. (23) takes the same form of Eq. (3). For linear heating rate function, the general solution of Eq. (23) is given by Eq. (24):

$$I = noS\exp\left(\frac{-E}{KT}\right)\exp\left(\frac{-S}{\beta}\int^{T}\exp\left(\frac{-E}{KT}\right)\mathrm{d}T\right) \tag{24}$$

Then from Eq. (22) with Randall and Wilkins [14] assumptions, we reached the first kinetics order equation.

The abovementioned second kinetics order, resulting from different assumptions associated with Eq. (22). In one set of assumptions, one can take $n(t) = m(t)$ which is not very different from the parametric equality $n_o = m_o$ once the assumption $n_c \ll n$ is made.

In addition, we have to assert the retraping dominates [15]

$$A_n(N-n) \gg mA_m \tag{25}$$

We also suppose that the trap is far from being saturated, i.e., the retrapping duration.

$$n \ll N \tag{26}$$

Then, from Eq. (26) one can write:

$$\frac{mA_m}{mA_m + A_n(N-n)} \cong \frac{mA_m}{A_n(N-n)} \tag{27}$$

Using the condition of Eq. (26) in Eq. (27) one gets:

$$\frac{mA_m}{mA_m + A_n(N-n)} \cong \frac{mA_m}{A_n(N-n)} \cong \frac{mA_m}{NA_n} \tag{28}$$

and since we have assumed that $n(t) = m(t)$,

$$\frac{mA_m}{mA_m + A_n(N-n)} \cong \frac{mA_m}{A_n(N-n)} \cong \frac{mA_m}{NA_n} \cong \frac{nA_m}{NA_n} \tag{29}$$

Then, apply to Eq. (22), we get:

$$I = \frac{-\mathrm{d}m}{\mathrm{d}t} = \frac{A_m S}{A_n N} n^2 \exp\left(\frac{-E}{KT}\right) \tag{30}$$

Alternatively, one can assume, in addition to the concentration equality, that $A_n = A_m$ [18] which yields:

$$I = \frac{-dm}{dt} = \frac{S}{N} n^2 \, exp\left(\frac{-E}{KT}\right) \tag{31}$$

Then Eq. (22) takes the same form of Eq. (8) which is found by Gralick and Gibson [15]. Where Eq. (30) sums up both these possibilities by employing the parameter S' ($m^3 s^{-1}$), the pre-exponential factor that replaces $\frac{A_m S}{A_n N}$ in one case and S/N in the other. The solution of Eq. (30) is given by Eq. (32)

$$I = S' no^2 \exp\left(\frac{-E}{KT}\right) \left\{1 + \left(\frac{S'}{\beta}\right) no \int^T \exp\left(\frac{-E}{KT}\right) \mathrm{d}T\right\}^{-2} \tag{32}$$

It should be emphasized that two cases discussed so far, namely first and second kinetics order, are only special cases in a sense, extreme cases and the general case described by equations Eq. (17) through Eq. (19) may be neither first nor second order even if the simplifying conditions of Eq. (21) are assumed to be general. The resulting Eq. (20) consists of many intermediate cases that do not have a distinct kinetics order. Although, some researchers still attempt to determine for every TL peak a first or second kinetics order [25].

Several attempts [16, 26] have been made to add a third parameter to the two basic ones, the activation energy E and the pre-exponential constant S' (or S), all the attempts extend the "order parameter" implied when talking about first or second-order peak. The order parameters considered so far as a discrete magnitude assuming the value of 1 and 2 can be extended to be a continuous parameter. It is to be noted, however, that the addition of a third parameter is in principle one step in the right direction since the general treatment should include eight parameters (E, S, A_m, A_n, N, n_o, m_o, n_c). The best-known way of including the third parameter is that of general kinetics order, b, according to which one can assume that the glow peak is governed by [25].

$$I = \frac{-\mathrm{d}n}{\mathrm{d}t} = S' n^b \exp\left(\frac{-E}{KT}\right) \tag{33}$$

The kinetics order, b, is normally considered to be between 1 and 2, but it can occasionally exceed this range [13]. The rationale behind writing Eq. (33) is as follows: it is readily seen that a first-order peak is asymmetric, where a second order peak is nearly symmetric. Following Halperin and Braner [19] and Chen [16] we can define the symmetry factor μ_g as:

$$\mu_g = \frac{\delta}{\omega} \tag{34}$$

where $\delta = T_2 - T_m, \omega = T_2 - T_1$ as it is shown in **Figure** 7, and T_1 and T_2 are the low and high temperatures on half- maximum intensity, respectively. It has been

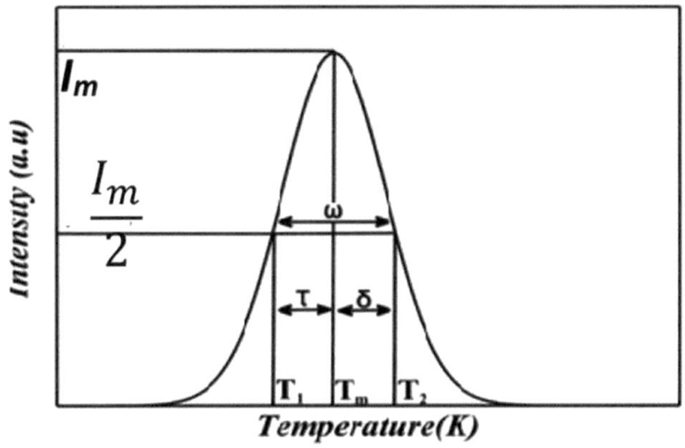

Figure 7.
Parameters used in the calculation of the symmetry factor.

shown [16] that for the first kinetics order, $\mu_g \cong 0.42$ and the second kinetics order, $\mu_g \cong 0.52$.

Of course, intermediate symmetries represented by different values of μ_g are found and the simplest way to present them by taking $1 < b < 2$ in Eq. (30). Chen [16] has shown that μ_g changes from 0.42 to 0.52 as b increasing from 1 to 2. The solution of Eq. (33) for linear heating rate β, is given by:

$$I = Sn_o\, exp\left(\frac{-E}{KT}\right)\left[\frac{(b-1)S}{\beta}\int_{T_o}^{T} exp\left(\frac{-E}{KT}\right)dt + 1\right]^{\frac{-b}{b-1}} \tag{35}$$

where $S = S'n_o^{b-1}$. Eq. (35) represents glow peak intensity according to the general kinetics order.

A few words of caution are in order with respect to this treatment. First, although Eq. (34) has been shown to quite accurately described measured TL peaks [27, 28], it is to be noted that in most cases it is only an empirical presentation and is not based on the three differential equations [Eqs. (17) up to (19)], seem to be more physically significant. However, the general order case is still important because it can handle intermediate circumstances and smooth the first and second-order cases as b_1 and b_2, respectively.

3.1.3 General-order kinetics

May and Partridge supposed the empirical equation that has been suggested to explain the thermoluminescence glow peak if the first or second-order kinetics do not describe the glow peak. The equation is namely the general- order kinetics and written by:

$$I = \frac{n_0\, s''\exp\left(-E/KT\right)}{\left[1 + [(b-1)s''/\beta]\int_{T_0}^{T}\exp\left(-E/T'\,K\right)dT'\right]^{b(b-1)}}$$

Hence $s'' = s(Nn_0)$ is called the pre-exponential factor, b the order of kinetics and the rang supposed between 1 and 2 but sometimes this rang has able to be greater than those. The pre-exponential factor s'' is constant for given the dose, however, it differs with changing the absorbed dose with n_0.

3.1.4 Trap parameters evaluation techniques

3.1.4.1 Empirical methods

We can deduce that the higher the peak temperature T_m, the higher the activation energy Urbach [29], and Urbach [30] found empirically for KCl crystals:

$$E(eV) = \frac{T_m(K)}{500} \tag{36}$$

This can also be written as $E = 23KT_m$ and it differs according to the types of the sample. Halperin [19] deduced $E = 38\ KT_m$ for NACL samples, and Miller and Bube [31] arrived at $E = 39\ KT_m$ for LiF.

The maximum intensity of the peak, according to Randall and Wilkins [12, 13], occurs around the temperature where the electron escape probability is 1 s-1. As a result of Eq. (1), we have:

$$P = S\exp\left(\frac{-E}{KT_m}\right) = 1\underset{\rightarrow}{\text{yields}}E = KT_m\, ln\,(S) \tag{37}$$

3.1.4.2 Initial rise method

According to Eqs. (6), (12) and (3), we can say that at the start of the glow peak (initial rise region) the TL intensity is proportional to $\exp(-E/kT)$, irrespective of whether the first kinetics order is obeyed or not [32]. This temperature relationship persists until the quantity of trapped electrons is drastically reduced. Hence, by plotting Log (I) versus 1/T, the value of E can be obtained from the slope of the straight line obtained. As a result, using the equation: it is possible to calculate E without knowing the frequency factor S:

$$E = -K\frac{ln\,(I)}{\frac{1}{T}} \tag{38}$$

From Eq. (6), we see that when T is slightly greater than T_m, the argument of the second exponential is very small and therefore the value of the exponential function is close to unity and varies very slowly with temperature. The temperature dependence of $I(t)$ is therefore dominated by the first exponential function, however the second exponential function decreases with increasing temperature and at higher temperatures it decreases very rapidly [13].

Therefore, the range of the initial rise must be chosen in which the second exponential function has minimum influence on the TL intensity temperature dependence. Therefore, it is necessary to restrict the temperature range such that the TL intensity does not exceed one-tenth of the maximum intensity [32].

Between temperatures T_1 and T_2 (both $< T_m$) corresponding to values equal to $a_1 I_m$ and $a_2 I_m$ respectively as in **Figure 8**, where:

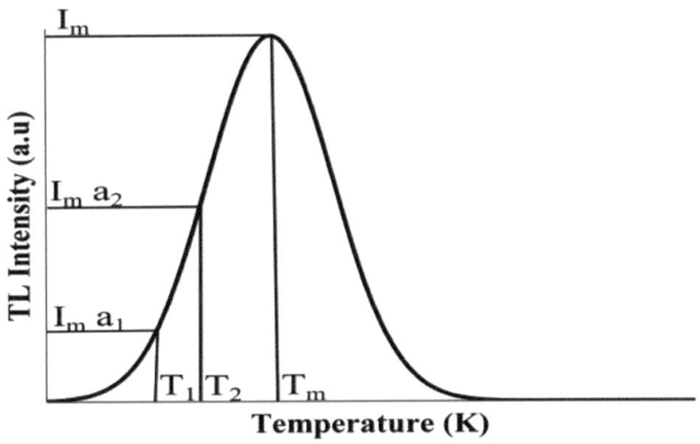

Figure 8.
Extracted parameters from "Christodoulides expression" are to correct the value of the activation energy evaluated by the initial rise method.

$$a_2 \leq 0.5, \frac{a_2}{a_1} \geq 5 \quad (39)$$

On the temperature scale, a series of points were taken at equal intervals and plotted as ln(I) versus (1/T). The value Ec can then be calculated from the slope of the straight line as the energy determined by the initial rise technique; this value is smaller than the real activation energy E by the amount that grows as a_1 and a_2 increase. Christodoulides [33] devised the following expression for the corrected energy E in terms of the measured values E_c, a_1, and a_2:

$$E = (1 + 0.74a_1 + 0.082a_2)E_c - \frac{(2a_1 + 0.22a_2)T_m}{11605} \quad (40)$$

The range of applicability of this equation is restricted by:

$$10 \ll \frac{E}{KT_m} \ll 100 \quad (41)$$

3.1.4.3 Peak shape method

Grossweiner [34] established the first peak shape approach for first-order peaks, writing:

$$E = 1.41K\frac{T_m T_1}{\tau} \quad (42)$$

Where: T_m is the temperature at the maximum intensity, T_1 is the temperature at the half of the maximum intensity in low-temperature side, $\tau = T_m - T_1$ as in **Figure 9**. Grossweiner used the coefficient 1.51, which was later [20] amended to 1.41. Lushchik [35] developed a method for evaluating the activation energy by utilizing the high-temperature half width $\delta = T_2 - T_1$ for first peaks he suggested:

$$E = \frac{KT_m^2}{\delta} \quad (43)$$

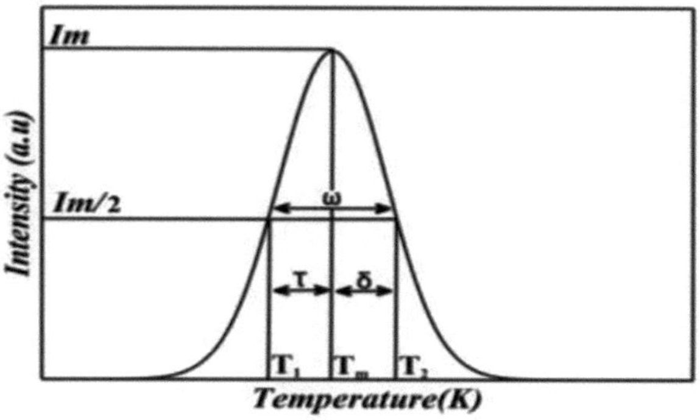

Figure 9.
Peak shape method used to calculate the activation energy.

and for second-order peak:

$$E = 2\frac{KT_m^2}{\delta} \tag{44}$$

Chen [16] improved these equations by adding a factor of 0.976 in front of the former and replacing the factor 2 by 1.71 in the latter.

Halperin and Braner [19] have derived their equations for both first [Eq. (45)] and second kinetics orders [Eq. (46)]:

$$E = 1.51K\frac{T_m^2}{\tau} - 3.16KT_m \tag{45}$$

$$E = 1.81K\frac{T_m^2}{\tau} - 4KT_m \tag{46}$$

Chen [16] managed to establish expressions for general kinetics order, which is dependent on the geometry factor of the glow peak which is defined by Eq. (35):

$$E = C_\varphi\frac{KT_m^2}{\varphi} - b_\varphi(2KT_m) \tag{47}$$

Where φ stands for τ, δ, ω and the values of C_φ and b_φ for the three methods are:

$$C_\tau = 3\left(\mu_g - 0.42\right) + 1.51 \tag{48}$$

$$C_\delta = 7.3\left(\mu_g - 0.42\right) + 0.976 \tag{49}$$

$$C_\omega = 10.2\left(\mu_g - 0.42\right) + 2.52 \tag{50}$$

$$b_\tau = 4.2\left(\mu_g - 0.42\right) + 1.58 \tag{51}$$

$$b_\omega = 1 \tag{52}$$

$$b_\delta = 0 \tag{53}$$

where μ_g is geometrical shape factor that equal $\frac{\delta}{\omega}$.

3.1.4.4 Various heating rates method

As mentioned above about Eq. (7), T_m changes with the heating rate β, writing Eq. (7) twice for heating rate β_1 and β_2 with maximum temperatures T_{m1} and T_{m2} we get [36]:

$$E = \frac{kT_{m1}T_{m2}}{T_{m2} - T_{m1}} \left\{ ln\left(\frac{\beta_1}{\beta_2}\right) - ln\left(\frac{T_{m1}^2}{T_{m2}^2}\right) \right\} \tag{54}$$

The activation energy that will be evaluated from Eq. (54) will be of course in accord with the first kinetics order only. However, Chen and Winer [37], Chen and Kirsh [38] showed that it can be used as a very good approximation for nonfirst-order cases as well.

The maximum condition, Eq. (16), can also take the following form:

$$-ln\left(\frac{\beta}{T_m^2}\right) = \frac{E}{K}\frac{1}{T_m} + ln\left(\frac{E}{SK}\right)$$

According to this equation, Hoogenstraaten [39] suggested using several heating rates, a plot of $\ln(\beta/T^2{}_m)$ vs. $(1/T_m)$ should yield a straight line of slope E/K, so that the activation energy is evaluated. Extrapolation to $1/T_m \rightarrow 0$ gives the value of ln(E/SK) from which the frequency factor is immediately found. It was shown that a plot of ln (I_m) versus $1/T_m$ for various heating rates usually yields a straight line too and the activation energy can be extracted similarly. It is to be noted from the theoretical point of view that β should be varied in as board a range as possible. However, this may cause various experimental difficulties. At very low heating rates, the maximum intensity will be low and in fact, the peak smeared, thus not allowing effective extraction of the experimental parameters. At high heating rates, a delay between the sample temperature and that of the measuring device impairs the temperature measurement. Moreover, temperature gradients within the sample usually occur at high heating rates which result in a smearing effect of a different kind. In practice, one should therefore compromise on a relatively narrow range of heating rates [10].

3.1.5 Three points method

A new technique was developed by Rasheedy [25], to evaluate the trap parameters from the measured glow curve according to the general kinetics order.

The behavior of a phosphor's TL intensity is determined by the following equation, [40], for generic kinetics order.

$$I = \frac{-dn}{dt} = \frac{n^b}{N^{b-1}} S \exp\left(\frac{-E}{KT}\right) \tag{55}$$

Where I is the intensity of the TL, n (cm^{-3}), is the electron concentration trapped at time t(s), N (cm^{-3}) is the traps concentration and K (eV/°K) is the Boltzman constant. Eq. (55) is more general than the two equations describing the first and second kinetics orders.

Eq. (55) is a modification of Eq. (33) in which the pre-exponential factor is defined as: $S' = \frac{S}{N^{b-1}}$ instead of $' = \frac{S}{n_o^{b-1}}$. The solution of Eq. (55) is given by Rasheedy [40]:

$$I = \frac{n_o S'' \exp\left(-\frac{E}{KT}\right)}{\left\{1 + \frac{(b-1)S''}{\beta} \int_{T_o}^{T} \exp\left(-\frac{E}{KT}\right) dT\right\}^{\frac{b}{b-1}}} \tag{56}$$

Where the pre-exponential factor $S'' = S(n_o/N)^{b-1}$ which is constant for a given dose but it varies with changes in the absorbed dose, i.e., with n_O.

This method is based on the proportional of the concentration of populated traps during the running of the TL to the area under the glow peak.

I_x is the TL intensity at temperature T_x at any portion of the glow peak as shown in **Figure 10**, then Eq. (55) becomes:

$$I_x = \frac{A_x^b}{N^{b-1}} S \exp\left(\frac{-E}{KT_x}\right) \tag{57}$$

Where A_x is the area under the glow peak between the temperatures T_x and T_f (the final temperature of glow peak). Similarly, we have:

$$I_y = \frac{I_x}{y} = \frac{A_y^b}{N^{b-1}} S \exp\left(\frac{-E}{KT_y}\right) \tag{58}$$

$$I_z = \frac{I_x}{z} = \frac{A_z^b}{N^{b-1}} S \exp\left(\frac{-E}{KT_z}\right) \tag{59}$$

Where I_y and I_z are the TL intensities at temperatures T_y and T_z, respectively.

From Eq. (57) and Eq. (58), we shall get

$$E = \left\{ ln\,(y) - bln\left(\frac{A_x}{A_y}\right) \right\} \left[\frac{KT_xT_y}{T_x - T_y}\right] \tag{60}$$

And from Eq. (57) and Eq. (59), we shall get:

$$E = \left\{ ln\,(z) - bln\left(\frac{A_x}{A_z}\right) \right\} \left[\frac{KT_xT_z}{T_x - T_z}\right] \tag{61}$$

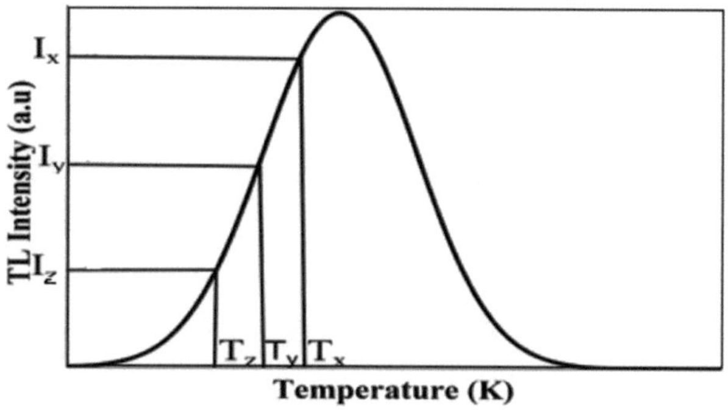

Figure 10.
Three points method used by Rasheedy [25] to investigate the equations used to calculate the trap parameters.

The order of kinetics, b can be obtained using Eqs. (60) and (61) which leads to:

$$b = \frac{T_y[T_x - T_z]\, ln\,(y) - T_z\left[T_x - T_y\right] ln\,(z)}{T_y[T_x - T_z]\, ln\left(\frac{A_x}{A_y}\right) - T_z\left[T_x - T_y\right] ln\left(\frac{A_x}{A_z}\right)} \tag{62}$$

Then, the order of kinetics b can be obtained from Eq. (62). Once the order of kinetics b is determined, the activation energy E(eV) can be determined by using Eq. (60) or Eq. (61).

Since, at $T = T_m$yields $\frac{\partial I}{\partial T} = 0$
$\rightarrow$

From Eq. (56) and using Eq. (59) leads to the following expression [41]:

$$S'' = \frac{\beta E\ \exp\left(\frac{E}{KT_m}\right)}{\left[bKT_m^2\right] - (b-1)E \exp\left(\frac{E}{KT_m}\right)\int_{T_o}^{T_m} \exp\left(-\frac{E}{KT}\right) dT} \tag{63}$$

A simple analytical method has been developed to obtain the relative value of n_o in the case of general kinetics order [41]:

$$n_o = \frac{I_m\, exp\left(\frac{E}{KT_m}\right)}{\acute{S}} \tag{64}$$

where T_m, and I_m can be obtained from the shape of the glow peak.

Thus, by calculating the kinetics order b, the activation energy E, and the initial trapped electrons number for many points that cover sufficient range on the glow peak, and taking the average value for each parameter, one can determine the trap parameters according to the general kinetics order.

3.1.6 Glow curve analysis (peak shape methods)

A review of the expression used in an intercomparison of glow curve analysis computer programs to evaluate TLD-100 glow curve is given in Ref. [42] where $I(T)$ is written in the following form:

$$I(T) = AS \exp\left(\frac{-E}{KT}\right)\left\{1 + \frac{S(b-1)}{\beta}\int_{T_o}^{T} \exp\left(\frac{-E}{KT}\right) \mathrm{d}t\right\}^{\frac{b}{1-b}} \tag{65}$$

where: A = area (counts); b = kinetics order; E = activation energy; I = intensity (counts per s, counts per K); S = frequency factor (s^{-1}).

On the other hand, Eq. (66) is based on first order kinetics which was used by Puchalska, [43], to develop glow-curve analysis software, in the following form:-

$$I(T) = I_m \exp\left(\frac{E}{KT_m} - \frac{E}{KT}\right) \exp\left(\frac{E}{KT_m}\left[\alpha\left(\frac{E}{KT_m}\right)\right] - \left(\frac{T}{T_m}\right)\left(\alpha\left(\frac{E}{KT}\right)\right) \exp\left(\frac{E}{KT_m} - \frac{E}{KT}\right)\right) \tag{66}$$

Where the parameter α is defined in Eq. (67) as

$$\alpha(x) = 1 - \frac{a_o + a_1x + a_2x^2 + a_3x^3 + x^4}{b_o + b_1x + b_2x^2 + b_3x^3 + x^4} \tag{67}$$

where the constants a_0, a_1 ... and b_0, b_1 ... are listed in the followings: -

$$a_o = 0.26777 b_o = 3.9584$$

$$a_1 = 8.63476 b_1 = 21.099653$$

$$a_2 = 18.05901 b_2 = 25.63295$$

$$a_3 = 8.573328 b_3 = 9.573322$$

Equation (68) will be used throughout our results which give better fitting to the resultant deconvoluted peaks. Different software was developed by Ratovonjanahary et al. [32], which uses the first kinetics order with an approximation of the second kinetics order. In this software the following equation was used:

$$I(T) = I_m \left\{ \exp 1 + \frac{E}{KT} \frac{T - T_m}{T_m} - \frac{T^2}{T_m^2} \exp\left(\frac{E}{KT} - \frac{T - T_m}{T_m}\right)(1 - \Delta) - \Delta_m \right\} \tag{68}$$

where,

$$\Delta = \frac{2KT}{E}, \Delta_m = \frac{2KT_m}{E}$$

Such a technique was also developed to analyze the glow curve using Eq. (53) by Rasheedy [41], which used the value of the trap parameters obtained by the three points method.

4. Modern clinical applications of TLD

TLD is widely used in various clinical fields for different purposes. The key reasons are undoubtedly their widespread availability, well-studied dosimetric characteristics, and applicability across a broad dose range. Imaging and Radiation Oncology Core-Houston IROC-H conducts remote dosimetry audits on MV photon and electron beams. IROC-H usually used integration between TLD-100 and other dosimetry system like nanoDot or diode systems for achieving the dose commissioning and calibrating dosimetry systems in an acrylic mini-phantom [44]. The failure rate was recorded in dose curves after modeling of the TPS (RayStation-Elekta Inc.) using phantom tests, which was not observed by patient-specific IMRT QA. Such failure was related to little changes in the MLC leaf-tip offset rather than leaf-tip width. Koger and his team [45] in IROC-H prosed four labeled TLD distributed in an anthropomorphic head-and-neck phantom for correcting such failure, (see **Figure 11**). It was utilized a 3D diode array were used in addition to assess the detectability of modeling mistakes [45].

Another crucial issue is to increase the staff's awareness about radiation safety and enhance radiation protection against unnecessary radiation doses. For such purpose, TLD-100 was recently used to validate occupational doses both inside and outside the nuclear

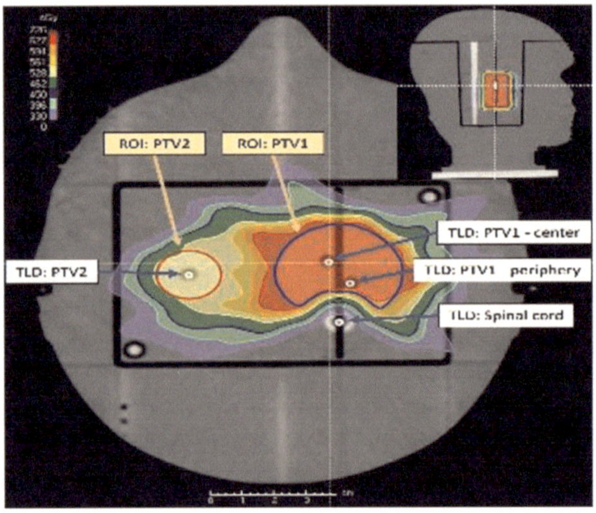

Figure 11.
TLDs were labeled in head-and-neck phantom at IROC-H [45].

medicine department, radiation protection purposes as well as the dose rate distribution around the positron emission tomography or computed tomography (PET/CT) [46].

Some recent studies were envisaged to see how the department compared to reports from other centers across the world in terms of the annual number of procedures and exposure limits, and to see if there was an opportunity for further radiation protection enhancements. As an example, personal TLD was calibrated to estimate the personal equivalent dose Hp (10) and Hp (0.07) at PET/CT. It was used for assessing the employee's exposure [47]. On the other hand, TLD rings personal dosimeters were worn by surgeons in their fingers through sentinel node biopsy procedure to measure personal doses Hp(10) and Hp(0.07), as well as ambiental dose for operating theater and during injection [48]. This will assure that personal equivalent doses are within the acceptable annual determined limits [49].

Other important recent TLD application in diagnostics is using an anthropomorphic phantom that modeled the reference person to get a conversion coefficient connecting dose area product (DAP) to effective patient dosage. They concluded that the effective dosage at the clinical dark-field radiography system, which generates both attenuation and dark-field pictures, is within the range of chest radiography standard dose values [50].

TLDs showed to be an excellent choice for skin dosimetry. Omojola et al. [51] used TLD in measurements of 3D skin dosimetry and verify their results using TPS planning verification at specific spots in the phantom. A full perspective of the dose distribution was achieved; however, they revealed that regions outside the PTV require special attention [52–54].

In addition, in the field of proton therapy, a novel tissue-equivalent TLD-sheet of manganese doped lithium triborate showed a valuable and effective dosimetry technique. It may also be a great in vivo skin dosimetry instrument for proton treatment due to its flexible and reusable properties. Despite the presence of significant energy dependences in the Bragg peak region, the response properties studied in this work, including as reproducibility, fading effects, dosage linearity and dose homogeneity are acceptable [55].

Monte Carlo (MC) simulation is considered a good tool to understand well the TLD [56]. Some algorithm methods based on MC as if pencil beam could be involved in accurate dose in MV radiotherapy calculations. It could be useful to calculate the spectrum inside the detector based on four categories primary photon and electrons and secondary photon and electrons [57].

On the other hand, Low-energy (100 keV) photons (x-rays and gamma) have been widely employed in biological research and medical applications for more than a century, including mammography, fluoroscopy, general radiography, computed tomography, and brachytherapy treatment, among others. The majority of electrons created by low photon energy beams have energies below 10 keV, according to research. The physical processes through which these low-energy electrons interact with matter, on the other hand, are still unknown. Furthermore, it is commonly thought that all energy put within a dosimeter-sensitive volume is converted into a response. However, this assumption could be inaccurate because some of the deposited energy could be utilized to build flaws or damages at the molecular and atomic levels [58].

The hybrid-functional density theory (H-DFT) has shown to be a promising tool for localizing secondary electrons within a dosimeter volume and calculating the energy spent on creating defects or colors centers, among other things, when it comes to the relationship between the energy deposited and the response of a dosimeter. Following that, the quantity of energy that can be truly turned into a dosimeter response following exposure to ionizing radiation would be more accurately determined.

5. Conclusion

The aim of this chapter is concerned with TLD materials, measurements and recent various applications in clinical and industrial fields. TL kinetics are also covered in details due to their importance in knowing traps parameters and band structure-related phenomena that are responsible for TL phenomena. Modern clinical applications of TLD are also covered like quality assurance purposes for proton, x-ray and gamma radiotherapy based on phantom tests. In addition, we shed spot on using TLD for recent accurate methods for skin dose evaluation under IMRT/VMAT radiotherapy. Special attention should be oriented to hybrid-functional density theory Monte Carlo simulation to model TL dosimeters. Recent studies proved a promising tool for localizing secondary electrons within a dosimeter volume and calculating the energy spent on creating defects or colors centers, among other things, when it comes to the relationship between the energy deposited and the response of a dosimeter. Such methods could give knowledge about misunderstanding behaviors of some TLD and could eliminate its disadvantages like missing TL signal or fading; angle and energy of incidence ionizing radiation. In general, the properties of TLD like its inexpensive cost and reusability; easily fabricated, lightweight to wear, readout is simple and can be automated, insensitive to humidity make it advantageous in different clinical and radiation safety applications.

Author details

Hossam Donya[1,2]

1 Faculty of Science, Department of Physics, King Abdulaziz University, Jeddah, Saudi Arabia

2 Faculty of Science, Physics Department, Menoufia University, Shebin El-Koom, Egypt

*Address all correspondence to: hdunia@kau.edu.sa

References

[1] Furuta Y, Tanaka S. Response of 6LiF and 7LiF thermoluminescence dosimeters to fast neutrons. Nuclear Instruments and Methods. 1972;**104**:36-374

[2] Moreira Ribeiro R, Souza-Santos D. Monte Carlo characterization of an individual albedo neutron monitor. Brazilian Journal of Radiation Sciences. 2021;**9**(2C):1-8. DOI: 10.15392/bjrs.v9i2C.1657

[3] El-Faramawy N, Chopra V, Rawash S, El-Hafez AA, Dhoble SJ. Response of TLD-600/TLD-700 and CR-39 to neutrons for medical dosimetry. Luminescence. 2021;**36**(5):1257-1264

[4] Gibson AB, Piesch E. Technical Reports Series No. 252. Neutron Monitoring for Radiological Protection. Vienna: International Atomic Energy Agency. 1985

[5] ICRU. Determination of Dose Equivalents Resulting from External Radiation Sources. In: Report 39, International Commission on Radiation Protection and Measurements. Bethesda, MD. 1985

[6] Seifert H, Dörschel B, Pawelke J, Hahn T. Comparison of calculated and measured neutron sensitivities of an electret albedo dosemeter. Radiation Protection Dosimetry. 1991;**37**(1):13-18

[7] Seifert H, Dorschel B, Pawelke J. Neutron dosimetry using an electret albedo dosimeter. In: 1991 Proceedings 7th International Symposium on Electrets (ISE 7). London: IEEE; 1991. pp. 753–758

[8] Fellinger J, Hahn T, Henniger J, Hübner K, Schmidt P. Fast neutron sensitivity of TL detectors using proton radiator techniques. Isotopenpraxis Isotopes in Environmental and Health Studies. 1991;**27**(7):342-346. DOI: 10.1080/10256019108622562

[9] El-Adawy A, Khaled NE, El-Sersy AR, Hussein A, Donya H. TL dosimetric properties of Li2O–B2O3 glasses for gamma dosimetry. Applied Radiation and Isotopes. 2010;**68**(6):1132-1136

[10] Polymeris GS, Çoskun S, Tsoutsoumanos E, Konstantinidis P, Aşlar E, Şahiner E, et al. Dose response features of quenched and reconstructed, TL and deconvolved OSL signals in BeO. Results in Physics. 2021;**25**:104222

[11] Begum M, Rahman AM, Abdul-Rashid HA, Yusoff Z, Nawi SN, Khandaker MU, et al. Photonic crystal fibre as a potential medium for radiotherapy dosimetry. Applied Radiation and Isotopes. 2021;**174**:109771

[12] Saray AA, Kaviani P, Shahbazi-Gahrouei D. Dosimetric characteristics of lithium triborate (LiB3O5) nanophosphor for medical applications. Radiation Measurements. 2021;**140**:106502

[13] Horowitz YS, Yossian D. Computerised glow curve deconvolution: Application to thermoluminescence dosimetry. Radiation Protection Dosimetry. 1995;**60**(1):1-114

[14] Randall JT, Wilkins MHF. Phosphorescence and electron traps II. The interpretation of long-period phosphorescence. Proceedings of the Royal Society of London. Series A. Mathematical and Physical Sciences. 1945;**184**(999):390-407

[15] Garlick GFJ, Wilkins MHF. Short period phosphorescence and electron traps. Proceedings of the Royal Society of

London. Series A. Mathematical and Physical Sciences. 1945;**184**(999):408-433

[16] Chen R. On the calculation of activation energies and frequency factors from glow curves. Journal of Applied Physics. 1969;**40**(2):570-585

[17] Wrzesińska A. Their production and thermoluminescence curves. Acta Physica Polonica. 1956;**15**:151

[18] Adirovitch EI. La formule de Becquerel et la loi élémentaire du déclin de la luminescence des phosphores cristallins. Journal de Physique et le Radium. 1956;**17**(8-9):705-707

[19] Halperin A, Braner AA, Ben-Zvi A, Kristianpoller N. Thermal activation energies in NaCl and KCl crystals. Physical Review. 1960;**117**(2):416

[20] Dussel GA, Bube RH. Theory of thermally stimulated conductivity in a previously photoexcited crystal. Physical Review. 1967;**155**(3):764

[21] Saunders IJ. The thermally stimulated luminescence and conductivity of insulators. Journal of Physics C: Solid State Physics. 1969; **2**(12):2181

[22] De Muer D. Development of a universal method for calculating the thermoluminescence parameters. Physica. 1970;**48**(1):1-12

[23] Bräunlich P, Kelly P, Fillard JP. Thermally stimulated luminescence and conductivity. In: Thermally Stimulated Relaxation in Solids. Berlin, Heidelberg: Springer; 1979. pp. 35-92

[24] Moharil SV. On the general-order kinetics in thermoluminescence. Physica Status Solidi A: Applications and Materials Science. 1982;**73**(2): 509-514

[25] Rasheedy MS. A new evaluation technique for analyzing the thermoluminescence glow curve and calculating the trap parameters. Thermochimica Acta. 2005;**429**(2): 143-147

[26] May CE, Partridge JA. Thermoluminescent kinetics of alpha-irradiated alkali halides. The Journal of Chemical Physics. 1964;**40**(5):1401-1409

[27] Prokić M. Analysis of the thermoluminescence glow curves of natural barite. Journal of Physics and Chemistry of Solids. 1977;**38**(6):617-622

[28] De Blasi C, Gallassini S, Manfredotti C, Micocci G, Ruggiero L, Tepore A. Trapping levels in PbI2. Solid State Communications. 1978;**25**(3):149-153

[29] Urbach R. Zur lumineszenz der alkalihalogenide. Sitzungsberichte Akad. der Wiss. Wien. 1930;**139**:363-372

[30] Urbach F. Storage and Release of Light by Phosphors. Vol. 115. New York: John Wiley and Sons; 1948

[31] Miller LD, Bube RH. Luminescence, trapping, and F centers in lithium fluoride crystals. Journal of Applied Physics. 1970;**41**(9):3687-3697

[32] Ratovonjanahary AJF, Raboanary R, Andriambololona R. Quartz Glow-Peaks Lifetime Analysis: TL Glow-Curve Deconvolution Functions for First Order of Kinetic Compared to Initial Rise Method. In: HEPMAD 04 Conference, Madagascar, 27 Sep-01 Oct. 2004

[33] Christodoulides C. Errors involved in the determination of activation energies in TL and TSDC by the initial rise method. Journal of Physics D: Applied Physics. 1985;**18**(8):1665

[34] Grossweiner LI. A note on the analysis of first-order glow curves. Journal of Applied Physics. 1953;**24**(10): 1306-1307

[35] Lushchik CB. The investigation of trapping centers in crystals by the method of thermal bleaching. Soviet Physics Jetp-USSR. 1956;**3**(3):390-399

[36] Booth AH. Calculation of electron trap depths from thermoluminescence maxima. Canadian Journal of Chemistry. 1954;**32**(2):214-215

[37] Chen R, Winer SAA. Effects of various heating rates on glow curves. Journal of Applied Physics. 1970;**41**(13): 5227-5232

[38] Chen R, Kirsh Y. Analysis of Thermally Stimulated Process. Oxford: Pergamon Press; 1981

[39] Hoogenstraaten W. Electron traps in zinc sulphide phosphors. Philips Research Reports. 1958;**13**:515-693

[40] Rasheedy MS. On the general-order kinetics of the thermoluminescence glow peak. Journal of Physics: Condensed Matter. 1993;**5**(5):633

[41] Rasheedy MSRMS. A complete system for obtaining the trap parameters of thermoluminescence glow peak. Japanese Journal of Applied Physics. 1996;**35**(2R):634

[42] Bos AJJ, Piters TM, Gómez-Ros JM, Delgado A. An intercomparison of glow curve analysis computer programs: I. Synthetic glow curves. Radiation Protection Dosimetry. 1993;**47**(1-4): 473-477

[43] Puchalska M, Bilski P. GlowFit—A new tool for thermoluminescence glow-curve deconvolution. Radiation Measurements. 2006;**41**(6): 659-664

[44] Alvarez P, Kry SF, Stingo F, Followill D. TLD and OSLD dosimetry systems for remote audits of radiotherapy external beam calibration. Radiation Measurements. 2017;**106**: 412-415

[45] Koger B, Price R, Wang D, Toomeh D, Geneser S, Ford E. Impact of the MLC leaf-tip model in a commercial TPS: Dose calculation limitations and IROC-H phantom failures. Journal of Applied Clinical Medical Physics. 2020;**21**(2): 82-88

[46] Nilsson I, Himmelman J, Khan J, Dalmo J. The potential to use Tld measurements to validate the occupational radiation protection at the Department of Nuclear Medicine. Radiation Protection Dosimetry. 2021; **195**(3-4):355-362

[47] Pavičar B, Davidović J, Petrović B, Vuleta G, Trivić S, Šajinović V, et al. Nuclear medicine staff exposure to ionising radiation in 18F-FDG PET/CT practice: A preliminary retrospective study. Arhiv za Higijenu Rada i Toksikologiju. 2021;**72**(3):216-223

[48] Petrovic B, Vicko F, Radovanovic D, Samac J, Tot A, Radovanovic Z, et al. Occupational radiation dose of personnel involved in sentinel node biopsy procedure. Journal of Medical Physics. 2021;**91**:117-120

[49] Ali W, Sulieman A, Tamam N, Boshara N, Aldhebaib A, Alkhorayef M, et al. Estimation of patients organ doses and staff exposure during bone scan examination. Radiation Physics and Chemistry. 2021;**188**:109693

[50] Frank M, Urban T, Willer K, Noichl W, De Marco F, Schick R, et al.

Dosimetry on first clinical dark-field chest radiography. Medical Physics. 2021;**48**(10):6152-6159

[51] Omojola AD, Akpochafor MO, Adeneye SO, Akala IO, Agboje AA. Chest X-rays of newborns in a medical facility: Variation between the entrance skin dose measurements using the indirect and direct methods for clinical dose audit. Japanese Journal of Radiology. 2021;**40**(2):1-7

[52] Moradi F, Khandaker MU, Mahdiraji GA, Ung NM, Bradley DA. Dose mapping inside a gamma irradiator measured with doped silica fibre dosimetry and Monte Carlo simulation. Radiation Physics and Chemistry. 2017; **140**:107-111

[53] Moradi F, Ung NM, Mahdiraji GA, Khandaker MU, Entezam A, See MH, et al. Angular dependence of optical fibre thermoluminescent dosimeters irradiated using kilo- and megavoltage X-rays. Radiation Physics and Chemistry. 2017;**135**:4-10

[54] Moradi F, Mahdiraji GA, Dermosesian E, Khandaker MU, Ung NM, Mahamd Adikan FR, et al. Influence of dose history on thermoluminescence response of Ge-doped silica optical fibre dosimeters. Radiation Physics and Chemistry. 2017; **134**:62-70

[55] Kato T, Sagara T, Komori S, Kato R, Takeuchi A, Narita Y. Dosimetric properties of a newly developed thermoluminescent sheet-type dosimeter for clinical proton beams. Journal of Applied Clinical Medical Physics. 2021;**22**(4):158-165

[56] Donya H, Seniwal B, Darwesh R, Fonseca TC. Prospective Monte Carlo simulation for choosing high efficient detectors for small-field dosimetry. In: Theory, Application, and Implementation of Monte Carlo Method in Science and Technology. London: IntechOpen; 2019

[57] Donya H. Pencil-beam fluence evaluation based on Monte Carlo simulations algorithm of high energetic treatment photons. Journal of Medical Signals and Sensors. 2018;**8**(2):81

[58] Massillon-JL G. Future directions on low-energy radiation dosimetry. Scientific Reports. 2021;**11**:10569. DOI: 10.1038/s41598-021-90152-3

Chapter 6

Comparative Dosimetric Study between ^{60}Co and ^{192}Ir BEBIG High Dose Rate Sources, Used in Brachytherapy, Using Monte Carlo N-Particle Extended

Said Elboukhari, Khalid Yamni, Hmad Ouabi, Taoufiq Bouassa and Lahcen Ait Mlouk

Abstract

The purpose from this work is the investigation for dosimetric parameters of the two new BEBIG sources, ^{60}Co and ^{192}Ir used in high-dose-rate brachytherapy. According to the full report of AAPM and ESTRO; air-kerma strength, dose rate constant, radial dose function, and 2D along & away dose rates tables were calculated. Moreover, a comparison was made between the calculated dosimetric parameters for the HDR sources simulated in this study. We used the MCN-PX to investigate the dosimetric parameters of both sources. The geometry of each source was defined in the input program of MCNPX, and each simulation was performed with an appropriate number of particle histories to get an acceptable Type A statistical uncertainty. The results obtained were tabulated and presented in graphical format; these results show a good agreement with other previous studies. The comparison made between the two simulated sources in this work shows a minor difference observed in the generated 2D along & away tables for complementing the commissioning of these sources within a TPS. This difference is considered negligible by the clinical specialists.

Keywords: Co-60 versus Ir-192, Monte Carlo investigation, dosimetric comparison, HEBD working group

1. Introduction

The widespread sources in afterloading devices operated in high-dose-rate brachytherapy (HDR) are ^{60}Co and ^{192}Ir. This work aimed to investigate the dosimetric parameters for both HDR sources manufactured by BEBIG (Eckert & Ziegler BEBIG GmbH, Germany), ^{60}Co model: Co0.A86 and ^{192}Ir model: GI192M11, used in HDR brachytherapy. According to the TG-43 U1 and HEBD Working Group Report, recommendations for high-energy photons emitting brachytherapy sources [1] were provided. The dosimetric parameters were calculated; The air-kerma strength, dose rate constant, radial dose function, and the 2D along & away dose rate table in Cartesian coordinates are calculated for both new BEBIG sources, except the 2D anisotropy function.

Several studies were made for the HDR brachytherapy sources with different geometries and nuclides; we have cited some of them in this work. Varieties of Monte Carlo codes have been used to investigate the HDR brachytherapy sources. The BEBIG Co0.A86 was investigated using Geant4 by Granero et al., (2007) [2], PENELOPE used by Guerrero et al., (2014) [3], a study was made by Anwarul et al., (2012) using the Monte Carlo code EGSnrc [4], and H. Badry et al., (2018) used EGS5 for simulation of the same source model [5].

For the ^{192}Ir model: GI192M11, a study was made by Perez-Calatayud et al., (2012) using the Geant3 Monte Carlo code [1], Geant4 was used by Granero et al., (2005) for the same source model [6]. The comparison was also made in the case of radial dose function with the results obtained for the source model BEBIG Ir2.A852 simulated by Granero et al., (2008) [7] and Belousov et al., (2014) [8]. The obtained results in this study were in good coherence with the published data. Monte Carlo simulations were provided following the records cited in the report of the research committee Task Group 268 from AAPM [9]. MCNPX code was already used in some previous studies we cited the use of the version: 2.4 by Alizadeh et al., (2015) for the HDR ^{192}Ir source Flexisource model [10]. Also, we have investigated the dosimetric parameters of the same ^{60}Co source in our previous study Elboukhari et al., (2020) using the version 2.7 of the code Monte Carlo N-Particles eXtended (MCNPX) [11], this new version of the code operates the new updated tables of cross sections from ENDF/B-VII.1 data. MCNPX is a general-purpose three-dimensional simulation tool providing the transports of 37 different particle types for criticality, dosimetry, shielding, detector response, and many other applications. On the contrary of previous MCNPX Monte Carlo codes, the version used in this work of MCNPX provided a high precision, and the uncertainties depending on cross section tables are considered negligible.

To evaluate the difference between the two sources simulated in this study within a clinical use, we have generated the 2D along & away tables for complementing the commissioning of these sources within a clinical treatment planning system. A minor difference was observed in the generated along & away dose rates for the range of distances considered in this work. These results could help in the choice of the appropriate nuclide to use in the treatment regarding operation costs and frequency for source change, especially for developing countries such as in North Africa. Also, different studies were performed concerning the clinic practice. A study of M. Andrassy et al., (2012) concerned the behavior in the treated volume [12]. In addition, the studies of Venselaar et al., (1996) and Candela et al., (2013) mentioned that the behavior of the two nuclides at shorter distances from the treated volume is different from that at larger distances [13, 14]. This result is also mentioned in the study of Strohmaier and Zwierzchowski in 2011 [15].

2. Materials and methods

In this study, Monte Carlo simulation for HDR brachytherapy sources was performed following the recommendations of the American Association of Physicists in Medicine (AAPM) and the European Society for Radiotherapy and Oncology (ESTRO) in the HEBD working group report [1]. The formula proposed for 2D dose rates is:

$$\dot{D}(\mathrm{r},\theta) = \mathrm{S_k}\Lambda\frac{G(r,\theta)}{G(r0,\theta0)}\mathrm{g_L(r)F}(\mathrm{r},\theta) \tag{1}$$

Where:

- -$\dot{D}$(r, θ) is the dose rate in water at the distance r in centimeters from a line source,;
- θ the polar angle specifying the point of interest;
- S_K the air-kerma strength in units of cGy cm^2 h^{-1};
- Λ the dose rate constant expressed in cGy h^{-1} U^{-1};
- $\frac{G(r,\theta)}{G(r0,\theta0)}$ is the geometry factor with the reference point (r_0 = 1 cm and θ_0 = 90);
- $g_L(r)$ the radial dose function (L = 3.5 mm for both of the simulated sources in this work, Co0.A86 and GI192M11);
- F(r,θ) is the 2D anisotropy function.

3. Sources descriptions and geometries

3.1 ^{60}Co HDR source

The ^{60}Co HDR source (model Co0.A86, manufactured by BEBIG) was simulated in this work, and all the comparisons were made for the same source model. It is composed of homogeny cobalt 60 cylindrical core with L = 3.5 mm (length) and 0.5 mm in diameter (density = 8.09 gcm^{-3}). The active core is surrounded by air shell and encapsulated in a stainless steel cylindrical capsule with 0.15 mm thickness and 1 mm for the external diameter. We considered 0.9 mm and 2 mm for the source cable diameter and length, respectively, **Figure 1a**. The activity of the cobalt source used in this work was A_0 = 81.56 GBq, and the cobalt 60 half-life is $t_{1/2}$ = 5.27 years. The density used for the stainless steel is 8.03 gcm^{-3} for both the capsule and the source cable [2].

3.2 ^{192}Ir HDR source

For the ^{192}Ir HDR source, model GI192M11, manufactured by E & Z BEBIG, was simulated in this work, the comparisons with the published data included: Ir2.A852, Flexisource, and GammaMed models. The BEBIG GI192M11 simulated in this study was composed of homogeny iridium 192 cylindrical core with L = 3.5 mm (length) and 0.6 mm in diameter, the density of iridium used in this work was 22.56 gcm^{-3}. The active core is surrounded by an air shell and then encapsulated in a stainless steel

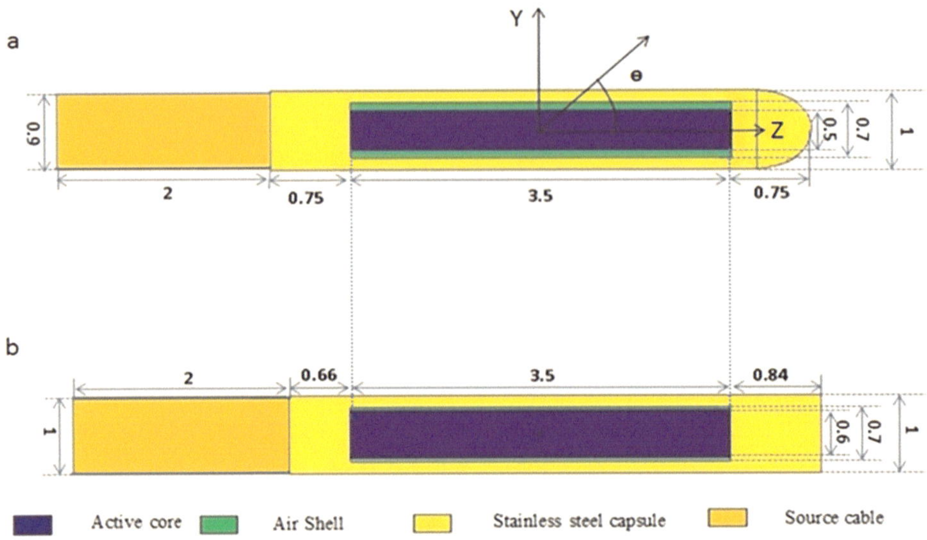

Figure 1.
(a) Schematic representation for the ^{60}Co source (model: Co0.A86). (b) Schematic representation for the ^{192}Ir source (model: GI192M11) (dimensions in mm).

cylindrical capsule of 0.15 mm thick and 1 mm for the external diameter, The source cable length and diameter used in this study, we consider, are 2 mm and 1 mm, respectively, **Figure 1b**; the activity of the iridium source used in this work was A_0 = 370GBq, and the iridium 192 half-life is $t_{1/2}$ = 73.81 days. The density used for the stainless steel is the same as used for the ^{60}Co (Co0.A86) source.

4. Monte Carlo calculations

For this work, we use the MCNPX version: 2.70 (license: C00810MNYCP) originally developed in the Los Alamos laboratory (Radiation Safety Information Computer Center, US). With the visual Editor VisedX_24E, this edition includes the package MCNP6.1/MCNP5–1.60/MCNPX-2.7.0. In addition, different tools were used for geometry modeling, particles transport, and 3D viewing of the defined geometry for the source and detectors with a dynamic model of simulation. To define the dosimetric parameters, MCNPX has different tallies to estimate each type of calculation. We consider the ^{60}Co source used this study composed of two gamma energies: 1.173 MeV and 1.332 MeV. The radiation spectrum of the ^{192}Ir source used was obtained from the database of (National Nuclear Data Center) neglecting the β spectrum for both of the simulated sources, since its contribution to the dose rate distribution is negligible due to the encapsulation [6, 16]. The Monte Carlo code fulfills all the recommendations of the report, "Dosimetric prerequisites for routine clinical use of photons emitting brachytherapy sources with average energy higher than 50 keV" of the AAPM and ESTRO. The following compilation options were used: CHEAP, DEC, PLOT, MCPLOT, GKSSIM, XS64, CEM, INCL, HISTP, MESHTAL, RADIOG, and SPABI. The physic models of MCNPX used in this study operating the new updated photons and electrons, the photons cross sections libraries mcplib02 and mcplib84 updated from mcplib04 photon Compton broadening data for MCNP5 [17], and the el03 for electrons.

The spectrum of gamma rays used in the simulations was obtained from the (National Nuclear Data Center) [18]; we use a cutoff energy of 10 keV for both of photons and electrons. Up to 2×10^9 photon histories were simulated in this study using an Intel® Xeon (R) CPU E5620@2.40GH × 16, HP-Z600 work station. No technique of variance reduction was used. To calculate the 2D along & away in water, the source was located in the center of a spherical phantom 40 cm in radius; acts like an unbounded phantom up to 20 cm from the source center for both ^{60}Co and ^{192}Ir sources. The density for the liquid water was 0.998 gcm^{-3} at 22°C according to the HEBD Working Group report. The coordinate axes used are shown in **Figure 1**. To obtain the radial dose function, and the along & away dose rate in the 2D Cartesian look-up table, we use a cylindrical rings system of 400*800 with 0.05 cm thick concentric to the longitudinal axis.

The high gamma energy of the ^{60}Co source takes electronic disequilibrium up to a distance of about 0.7 cm in water. Thus, we cannot approximate kerma by the dose in the near region to the source as in the case of ^{192}Ir. Consequently, the doses have been scored in distances near the source. The scored values for dose rate were included in the tables given in this study for the located points at distances where electronic disequilibrium exists. For distance greater than 1 cm from the source, to decrease the statistical uncertainty, the dose was approximate by the scored kerma; a previous study of Ballester et al., (2005) mentioned that the differences between dose and kerma are negligible at distances greater than 1 cm [19]. 10^9 Photon histories were

Element	**Medium**		
	Water (%)	**Air (%)**	**Stainless steel (%)**
H	11.010	0.073	—
C	—	0.012	0.03
N	—	75.032	0.01
O	88.900	—	—
Si	—	—	0.75
P	—	—	0.045
S	—	—	0.03
Ar	—	1.274	—
Cr	—	—	17.0
Mn	—	—	2.0
Fe	—	—	65.543
Ni	—	—	12.0
Mo	—	—	2.5
Co	—	—	—
Ir	—	—	—
Total mass percentage	99.910	76.391	99.908
Density (g/cm^3)	0.998	0.012	8.03

Table 1.
Elemental composition used in this study by mass percentage for: sources and water phantom, (international commission on radiation units and measurements, ICRU report 44, 1989) [20].

Monte Carlo study	S_K/A (*$10^{-7}UBq^{-1}$)	Λ ($cGyh^{-1}U^{-1}$)	Source model
^{60}Co			
Geant4 (Granero et al. 2007)	—	1.087 ± 0.011	BEBIG Co0.A86
EGSnrc (Anwarul et al. 2012)	3.039 ± 0.004	1.097 ± 0.001	BEBIG Co0.A86
PENELOPE (Guerrero et al. 2014)	3.046 ± 0.007	1.094 ± 0.003	BEBIG Co0.A86
EGS5 (H.Badry et al. 2018)	3.042 ± 0.007	1.092 ± 0.008	BEBIG Co0.A86
MCNPX (This work)	3.030 ± 0.002	1.092 ± 0.001	BEBIG Co0.A86
^{192}Ir			
Geant3 (Perez Calatayud et al. 2012)	1.091	1110	BEBIG GI192M11
Geant4 (Granero et al. 2005)	—	1.108 ± 0.003	BEBIG GI192M11
MCNPX (This work)	1.092 ± 0,004	1.108 ± 0.004	BEBIG GI192M11

Table 2.
Per unite source activity and Λ, obtained with MCNPX, compared with the values obtained in other previous studies.

simulated to obtain dose rate values in the region of electronic disequilibrium and 2.10^9 photon histories to score kerma for the ^{60}Co source. 10^9 photon histories were used to estimate kerma for the ^{192}Ir source.

To investigate the air-kerma strength, we kept the source in the center of a cubic phantom with $5 \times 5 \times 5\ m^3$ in dimensions. Then, the air-kerma was scored at 1 m in the transversal axis of the source using 1 mm thickness cylindrical rings, concentrated from distance 99.5 cm to 100.5 cm, filled with air, with relative humidity of 40% and mass density 0.001205 g cm^{-3}. In addition to that, to avoid the correction for photon attenuation and scatter in air, we have considered outside the scoring cells filled with vacuum. Elemental composition of materials used in this simulation is shown in **Table 1**, taken from the (ICRU 44 report) [20].

The dose rate constant was calculated using Eq. (2), by dividing the scored value of dose in a cubic voxel, with $0.1 \times 0.1 \times 0.1\ mm^3$ in dimensions by the air-kerma strength. Therefore, the scoring zone located in 1 cm from the active core center in the transversal axis (Y-axis), in a spherical phantom of 40 cm in radius filled with water.

$$\Lambda = \frac{\dot{D}(r,\theta)}{S_k} \tag{2}$$

The values of and Λ were compared with the published data and presented in **Table 2**.

5. Air kerma strength

The TG-43 formalism and the full report for the HEBD Working Group of the AAPM and ESTRO recommend for HDR brachytherapy specifying photon-emitting sources in terms of the air-kerma strength S_K, taking into account correction for attenuation and scattering in air. The relation between S_K and Kair is given by Eq. (3) [21]:

$$S_K = Kair_{dref} \times d_{ref}^2 \tag{3}$$

Where the reference air-kerma rate is defined at d_{ref} = 1 m.

The air-kerma per source photon depends to the photon fluence by the equation:

$$Kair = 1.602\,10^{-10} * \int_{Emin}^{Emax} \phi(E)E\left(\frac{\mu en(E)}{\rho}\right)dE \tag{4}$$

Where Kair is air kerma per source photon in Gy, the factor 1.602 10^{-10} converted the result from MeV g^{-1} into Gy, photon fluence (cm^{-2}) at the energy E (MeV) per initial source photon at the distance d, and the mass-energy absorption coefficient ($cm^{-2}\,g^{-1}$) at the energy E [22].

To obtain the total air-kerma, we use the following Eq. (8) [22].

$$\dot{Kair} = 1.602\,10^{-10} * \sum_{Emin}^{Emax} \phi(Ei)Ei\left(\frac{\mu en(Ei)}{\rho}\right)\Delta E \tag{5}$$

The total air-kerma per incident photon, *Ei* the midpoint for an energy bin, *ΔE* the bin size in MeV, for this study we use the photon fluence spectrum in 10 keV intervals. Thus, we introduce the Eq. (5) by using the MCNPX F6 tally, which is a track-length estimator [23], providing results in (MeV/g) [10, 24], converted into Gy by using the appropriate FM card tally multiplier (FM = 1.60210^{-10}). The composition for air is taken from the tables of X-ray mass attenuation coefficients and mass energy-absorption coefficients (NIST) [25]. The HEBD recommended a short-hand notation for the air-kerma strength: 1 U = 1μGym2 h^{-1} = 1cGycm^2 h^{-1}.Then to calculate the air-kerma strength per unit of source activity in (Gym^2s^{-1} Bq^{-1}), we use Eq. (6) below:

$$\frac{Sk}{A} = \dot{Kair}(d_{\mathrm{ref}})d_{\mathrm{ref}^2}\mathrm{N} \tag{6}$$

Where A is the source activity (Bq) and N the number of photons per decay, considered equal 2 for the ^{60}Co source, and 2.21 for the ^{192}Ir source.

6. Radial dose function

The radial dose function $g_L(r)$ described in the protocol of the (HEBD) takes into account scattering and absorption in the transversal axis of the source; it was calculated in a spherical phantom filled with water using concentric cylindrical rings to the longitudinal axis, with 0.05 cm thickness for the ranging distance from 0.25 cm to 20 cm for both of the simulated sources in this study. The results obtained are presented in **Table 3**.

7. Along & away absorbed dose

The along & away absorbed dose rates were investigated for the ranging distance from 0.25 cm to 7 cm in the transversal axis and from 0 cm to ±7 cm in the longitudinal axis. The 2D along & away was compared with the published data. The results

Radial distance r (cm)	$g_L(r)$	
	BEBIG Co0.A86 ^{60}Co	BEBIG GI192M11 ^{192}Ir
0.25	1.0705 ± 0.0041	0.9943 ± 0.0002
0.5	1.0221 ± 0.0046	0.9987 ± 0.0002
0.75	0.9938 ± 0.0048	0.9990 ± 0.0003
1	1 ± 0.0002	1 ± 0.0003
1.5	0.9926 ± 0.0002	1.0037 ± 0.0004
2	0.9874 ± 0.0002	1.0090 ± 0.0004
3	0.9689 ± 0.0003	1.0086 ± 0.0005
4	0.9539 ± 0.0003	1.0088 ± 0.0006
5	0.9378 ± 0.0003	1.0052 ± 0.0007
6	0.9205 ± 0.0003	0.9974 ± 0.0007
7	0.9035 ± 0.0003	0.9874 ± 0.0008
8	0.8867 ± 0.0003	0.9748 ± 0.0008
9	0.8683 ± 0.0003	0.9593 ± 0.0009
10	0.8513 ± 0.0003	0.9418 ± 0.0009
12	0.8156 ± 0.0003	0.9011 ± 0.0010
15	0.7593 ± 0.0003	0.8272 ± 0.0011
20	0.6628 ± 0.0003	0.6870 ± 0.0014

Table 3.
Radial dose function obtained for ^{60}Co and ^{192}Ir using MCNPX in a water.

are tabulated in the form recommended by the HEBD Working Group report [1], **Tables 4** and 5, respectively, for ^{60}Co and ^{192}Ir.

8. Uncertainties

The uncertainties evaluated in this study are the type A (k = 1) statistical uncertainty contribution dependent on the Monte Carlo technique. No technique of the variance reduction was used as mentioned before. All MCNPX results are normalized to be per initial particle history printed in the output with an additional number beside, which is the estimated statistical uncertainty. In this work statistical uncertainties are less than 0.8% and 0.4% type A uncertainty (k = 1) respectively for ^{60}Co and ^{192}Ir, derived by considering the contribution of the different simulated parameters for both of the simulated sources. In addition to the contribution of the propagated uncertainty for both of radial dose function and the 2D along & away in the relative uncertainties of the MCNPX output tallies. For the cobalt source dose rates, uncertainties were calculated from the quadrature sum of uncertainties obtained for the dose scored in the near distance to the source, and the scored kerma for the distance where the electronic equilibrium is reached.

Type B uncertainties are difficult to evaluate because of different contributions such as uncertainties of the cross section and energy spectrum, uncertainties in the modeled geometry of the source, and uncertainties in the scoring dose and kerma

Distance along Z-axis(cm)	Distance away in Y-Axis (cm)										
	0.25	0.5	0.75	1	1.5	2	3	4	5	6	7
−7	0.0189 ± 0.0109	0.0172 ± 0.0103	0.0206 ± 0.0099	0.0187 ± 0.0026	0.0187 ± 0.0022	0.0182 ± 0.0021	0.0168 ± 0.0021	0.0149 ± 0.0021	0.0130 ± 0.0021	0.0112 ± 0.0020	0.0096 ± 0.0020
−6	0.0220 ± 0.0105	0.0213 ± 0.0097	0.0295 ± 0.0093	0.0259 ± 0.0024	0.0257 ± 0.0022	0.0247 ± 0.0021	0.0220 ± 0.0021	0.0190 ± 0.0021	0.0160 ± 0.0021	0.0134 ± 0.0020	0.0112 ± 0.0020
−5	0.0350 ± 0.0101	0.0406 ± 0.0091	0.0392 ± 0.0088	0.0381 ± 0.0023	0.0369 ± 0.0021	0.0350 ± 0.0021	0.0297 ± 0.0020	0.0245 ± 0.0020	0.0199 ± 0.0020	0.0160 ± 0.002	0.0130 ± 0.0020
−4	0.0668 ± 0.0091	0.0613 ± 0.0086	0.0608 ± 0.0085	0.0597 ± 0.0022	0.0564 ± 0.0021	0.0516 ± 0.0020	0.0412 ± 0.0020	0.0319 ± 0.0020	0.0245 ± 0.0020	0.0190 ± 0.0020	0.0150 ± 0.0020
−3	0.1092 ± 0.0087	0.1099 ± 0.0083	0.1044 ± 0.0081	0.1045 ± 0.0021	0.0935 ± 0.0020	0.0811 ± 0.0020	0.0581 ± 0.0020	0.0412 ± 0.0020	0.0300 ± 0.0020	0.0222 ± 0.0020	0.0169 ± 0.0020
−2	0.2578 ± 0.0081	0.2640 ± 0.0078	0.2515 ± 0.0075	0.2149 ± 0.0021	0.1716 ± 0.0020	0.1340 ± 0.0020	0.0811 ± 0.0020	0.0521 ± 0.0020	0.0355 ± 0.0020	0.0252 ± 0.0020	0.0187 ± 0.0020
−1.5	0.4886 ± 0.0075	0.4445 ± 0.0073	0.3886 ± 0.0070	0.3355 ± 0.0020	0.2403 ± 0.0020	0.1725 ± 0.0020	0.0942 ± 0.0020	0.0573 ± 0.0020	0.0378 ± 0.0020	0.0264 ± 0.0020	0.0194 ± 0.0020
−1	1.0378 ± 0.0072	0.8976 ± 0.0068	0.7078 ± 0.0066	0.5473 ± 0.0020	0.3355 ± 0.0020	0.2163 ± 0.0020	0.1064 ± 0.0020	0.0616 ± 0.0020	0.0397 ± 0.0020	0.0274 ± 0.0020	0.0199 ± 0.0020
−0.75	1.8579 ± 0.0068	1.3801 ± 0.0066	0.9897 ± 0.0064	0.7019 ± 0.0020	0.3878 ± 0.0020	0.2375 ± 0.0020	0.1114 ± 0.0020	0.0633 ± 0.0020	0.0404 ± 0.0020	0.0276 ± 0.0020	0.0201 ± 0.0020
−0.5	3.8827 ± 0.0060	2.3749 ± 0.0062	1.4126 ± 0.0063	0.8780 ± 0.0020	0.4354 ± 0.0020	0.2546 ± 0.0020	0.1152 ± 0.0020	0.0645 ± 0.0020	0.0409 ± 0.0020	0.0278 ± 0.0020	0.0202 ± 0.0020
−0.25	10.1078 ± 0.0047	3.6613 ± 0.0059	1.7678 ± 0.0061	1.0279 ± 0.0020	0.4711 ± 0.0020	0.2665 ± 0.0020	0.1178 ± 0.0020	0.0654 ± 0.0020	0.0412 ± 0.0020	0.0281 ± 0.0020	0.0203 ± 0.0020
0	16.9827 ± 0.0041	4.4647 ± 0.0046	1.9732 ± 0.0048	1.0898 ± 0.0020	0.4830 ± 0.0020	0.2713 ± 0.0020	0.1185 ± 0.0020	0.0657 ± 0.0020	0.0414 ± 0.0020	0.0281 ± 0.0020	0.0203 ± 0.0020
0.25	10.0821 ± 0.0044	3.6851 ± 0.0050	1.8212 ± 0.0053	1.0279 ± 0.0020	0.4711 ± 0.0020	0.2665 ± 0.0020	0.1175 ± 0.0020	0.0654 ± 0.0020	0.0412 ± 0.0020	0.0281 ± 0.0020	0.0203 ± 0.0020
0.5	4.0092 ± 0.0047	2.2930 ± 0.0054	1.3918 ± 0.0055	0.8780 ± 0.0020	0.4354 ± 0.0020	0.2546 ± 0.0020	0.1152 ± 0.0020	0.0647 ± 0.0020	0.0409 ± 0.0020	0.0278 ± 0.0020	0.0202 ± 0.0020
0.75	1.8878 ± 0.0059	1.3874 ± 0.0059	1.0309 ± 0.0057	0.7019 ± 0.0020	0.3878 ± 0.0020	0.2375 ± 0.0020	0.1114 ± 0.0020	0.0633 ± 0.0020	0.0404 ± 0.0020	0.0276 ± 0.0020	0.0201 ± 0.0020
1	1.0680 ± 0.0067	0.8909 ± 0.0064	0.7267 ± 0.0062	0.5473 ± 0.0020	0.3355 ± 0.0020	0.2165 ± 0.0020	0.1064 ± 0.0020	0.0616 ± 0.0020	0.0397 ± 0.0020	0.0274 ± 0.0020	0.0199 ± 0.0020
1.5	0.4672 ± 0.0071	0.4276 ± 0.0069	0.3945 ± 0.0066	0.3355 ± 0.0020	0.2403 ± 0.0020	0.1725 ± 0.0020	0.0942 ± 0.0020	0.0573 ± 0.0020	0.0378 ± 0.0020	0.0264 ± 0.0020	0.0194 ± 0.0020
2	0.2684 ± 0.0078	0.2406 ± 0.0072	0.2382 ± 0.0070	0.2146 ± 0.0021	0.1716 ± 0.0020	0.1340 ± 0.0020	0.0811 ± 0.0020	0.0521 ± 0.0020	0.0355 ± 0.0020	0.0252 ± 0.0020	0.0187 ± 0.0020
3	0.1211 ± 0.0083	0.1093 ± 0.0077	0.1053 ± 0.0074	0.1049 ± 0.0021	0.0937 ± 0.0021	0.0811 ± 0.0021	0.0578 ± 0.0020	0.0412 ± 0.0020	0.0300 ± 0.0020	0.0222 ± 0.0020	0.0169 ± 0.0020
4	0.0593 ± 0.0088	0.0630 ± 0.0081	0.0603 ± 0.0078	0.0602 ± 0.0022	0.0564 ± 0.0021	0.0519 ± 0.0021	0.0412 ± 0.0021	0.0319 ± 0.0021	0.0245 ± 0.0021	0.0191 ± 0.0020	0.0150 ± 0.0020
5	0.0315 ± 0.0093	0.0381 ± 0.0087	0.0379 ± 0.0081	0.0385 ± 0.0023	0.0371 ± 0.0021	0.0350 ± 0.0021	0.0297 ± 0.0021	0.0245 ± 0.0021	0.0198 ± 0.0021	0.0160 ± 0.0020	0.0130 ± 0.0020
6	0.0221 ± 0.0100	0.00295 ± 0.0093	0.0215 ± 0.0091	0.0264 ± 0.0024	0.0257 ± 0.0022	0.0247 ± 0.0022	0.0221 ± 0.0021	0.0190 ± 0.0021	0.0160 ± 0.0021	0.0134 ± 0.0020	0.0112 ± 0.0020
7	0.0176 ± 0.0104	0.0192 ± 0.0101	0.0187 ± 0.0098	0.0192 ± 0.0026	0.0189 ± 0.0022	0.0184 ± 0.0022	0.0168 ± 0.0021	0.0149 ± 0.0021	0.0130 ± 0.0021	0.0112 ± 0.0020	0.0096 ± 0.0020

Table 4.
Dose rate along & away per unit air-kerma strength ($cGyh^{-1}$ U^{-1}) in liquid water obtained for the BEBIG ^{60}Co source (model: Coo.A86), (+Z toward the source tip and -Z toward the delivery cable).

Distance along Z-axis (cm)	Distance away in Y-axis (cm)										
	0.25	0.5	0.75	1	1.5	2	3	4	5	6	7
−7	0.0158 ± 0.0047	0.016 ± 0.0033	0.0164 ± 0.0027	0.0170 ± 0.0023	0.0174 ± 0.0019	0.0176 ± 0.0016	0.0169 ± 0.0013	0.0155 ± 0.0012	0.0137 ± 0.0011	0.0120 ± 0.0011	0.0103 ± 0.0011
−6	0.0210 ± 0.0042	0.0218 ± 0.0029	0.0225 ± 0.0024	0.0232 ± 0.0020	0.0240 ± 0.0016	0.0239 ± 0.0014	0.0223 ± 0.0012	0.0198 ± 0.0011	0.0171 ± 0.0010	0.0144 ± 0.0010	0.0122 ± 0.0010
−5	0.0297 ± 0.0036	0.0311 ± 0.0025	0.0326 ± 0.0020	0.0338 ± 0.0017	0.0345 ± 0.0014	0.0340 ± 0.0012	0.0302 ± 0.0010	0.0256 ± 0.0010	0.0212 ± 0.0009	0.0173 ± 0.0009	0.0142 ± 0.0010
−4	0.0455 ± 0.0030	0.0488 ± 0.0020	0.0516 ± 0.0016	0.0533 ± 0.0014	0.0533 ± 0.0011	0.0507 ± 0.0010	0.0422 ± 0.0009	0.0335 ± 0.0009	0.0263 ± 0.0009	0.0207 ± 0.0009	0.0164 ± 0.0009
−3	0.0800 ± 0.0023	0.0884 ± 0.0015	0.0935 ± 0.0012	0.0944 ± 0.0011	0.0897 ± 0.0009	0.0805 ± 0.0008	0.0599 ± 0.0008	0.0437 ± 0.0008	0.0321 ± 0.0008	0.0242 ± 0.0008	0.0186 ± 0.0008
−2	0.1839 ± 0.0015	0.2069 ± 0.0010	0.2100 ± 0.0008	0.2007 ± 0.0007	0.1688 ± 0.0007	0.1353 ± 0.0006	0.0846 ± 0.0006	0.0554 ± 0.0007	0.0381 ± 0.0007	0.0275 ± 0.0008	0.0206 ± 0.0008
−1.5	0.3408 ± 0.0012	0.3736 ± 0.0008	0.3563 ± 0.0006	0.3201 ± 0.0006	0.2394 ± 0.0006	0.1756 ± 0.0006	0.0987 ± 0.0006	0.0610 ± 0.0006	0.0408 ± 0.0007	0.0288 ± 0.0007	0.0213 ± 0.0008
−1	0.8279 ± 0.0007	0.8053 ± 0.0005	0.6747 ± 0.0005	0.5376 ± 0.0005	0.3374 ± 0.0005	0.2220 ± 0.0005	0.1116 ± 0.0006	0.0658 ± 0.0006	0.0429 ± 0.0007	0.0299 ± 0.0007	0.0219 ± 0.0008
−0.75	1.5283 ± 0.0005	1.2924 ± 0.0004	0.9574 ± 0.0004	0.6973 ± 0.0004	0.3921 ± 0.0004	0.2441 ± 0.0005	0.1168 ± 0.0005	0.0675 ± 0.0006	0.0437 ± 0.0007	0.0303 ± 0.0007	0.0221 ± 0.0008
−0.5	3.4028 ± 0.0004	2.1718 ± 0.0003	1.3443 ± 0.0003	0.8780 ± 0.0004	0.4429 ± 0.0004	0.2626 ± 0.0005	0.1208 ± 0.0005	0.0689 ± 0.0006	0.0442 ± 0.0007	0.0305 ± 0.0007	0.0222 ± 0.0008
−0.25	9.1268 ± 0.0002	3.4770 ± 0.0003	1.7488 ± 0.0003	1.0360 ± 0.0003	0.4798 ± 0.0004	0.2750 ± 0.0004	0.1234 ± 0.0004	0.0696 ± 0.0005	0.0445 ± 0.0006	0.0307 ± 0.0007	0.0224 ± 0.0007
0	15.4135 ± 0.0002	4.2627 ± 0.0002	1.9391 ± 0.0003	1.1005 ± 0.0003	0.4937 ± 0.0004	0.2796 ± 0.0004	0.1244 ± 0.0005	0.0700 ± 0.0006	0.0446 ± 0.0007	0.0307 ± 0.0007	0.0224 ± 0.0008
0.25	9.1199 ± 0.0002	3.4777 ± 0.0003	1.7498 ± 0.0003	1.0352 ± 0.0003	0.4801 ± 0.0004	0.2750 ± 0.0004	0.1234 ± 0.0005	0.0697 ± 0.0006	0.0445 ± 0.0007	0.0307 ± 0.0007	0.0223 ± 0.0008
0.5	3.3981 ± 0.0004	2.1709 ± 0.0003	1.3443 ± 0.0003	0.8782 ± 0.0004	0.4430 ± 0.0004	0.2627 ± 0.0005	0.1209 ± 0.0005	0.0690 ± 0.0006	0.0442 ± 0.0007	0.0305 ± 0.0007	0.0223 ± 0.0008
0.75	1.5284 ± 0.0005	1.2915 ± 0.0004	0.9570 ± 0.0004	0.6970 ± 0.0004	0.3921 ± 0.0004	0.2442 ± 0.0005	0.1168 ± 0.0005	0.0675 ± 0.0006	0.0437 ± 0.0007	0.0303 ± 0.0007	0.0221 ± 0.0008
1	0.8355 ± 0.0007	0.8050 ± 0.0005	0.6741 ± 0.0005	0.5373 ± 0.0005	0.3373 ± 0.0005	0.2217 ± 0.0005	0.1114 ± 0.0006	0.0657 ± 0.0006	0.0429 ± 0.0007	0.0299 ± 0.0007	0.0219 ± 0.0008
1.5	0.3519 ± 0.0011	0.3735 ± 0.0008	0.3561 ± 0.0006	0.3198 ± 0.0006	0.2394 ± 0.0006	0.1755 ± 0.0006	0.0985 ± 0.0006	0.0610 ± 0.0006	0.0408 ± 0.0007	0.0289 ± 0.0007	0.0213 ± 0.0008
2	0.1931 ± 0.0015	0.2090 ± 0.0010	0.2101 ± 0.0008	0.2006 ± 0.0007	0.1687 ± 0.0007	0.1352 ± 0.0006	0.0847 ± 0.0006	0.0554 ± 0.0007	0.0381 ± 0.0007	0.0275 ± 0.0008	0.0205 ± 0.0008
3	0.0857 ± 0.0022	0.0910 ± 0.0015	0.0943 ± 0.0012	0.0947 ± 0.0011	0.0898 ± 0.0009	0.0805 ± 0.0008	0.0598 ± 0.0008	0.0436 ± 0.0008	0.0322 ± 0.0008	0.0242 ± 0.0008	0.0186 ± 0.0008
4	0.0488 ± 0.0029	0.0506 ± 0.0020	0.0527 ± 0.0016	0.0537 ± 0.0014	0.0534 ± 0.0011	0.0508 ± 0.0010	0.0422 ± 0.0009	0.0335 ± 0.0009	0.0263 ± 0.0009	0.0207 ± 0.0009	0.0164 ± 0.0009
5	0.0316 ± 0.0035	0.0325 ± 0.0024	0.0336 ± 0.0020	0.0342 ± 0.0017	0.0347 ± 0.0014	0.0340 ± 0.0012	0.0303 ± 0.0010	0.0256 ± 0.0010	0.0212 ± 0.0009	0.0174 ± 0.0009	0.0142 ± 0.0010
6	0.0222 ± 0.0041	0.0228 ± 0.0029	0.0232 ± 0.0023	0.0236 ± 0.0020	0.0241 ± 0.0016	0.0240 ± 0.0014	0.0223 ± 0.0012	0.0198 ± 0.0011	0.0170 ± 0.0010	0.0144 ± 0.0010	0.0122 ± 0.0010
7	0.0165 ± 0.0047	0.0168 ± 0.0033	0.0170 ± 0.0027	0.0174 ± 0.0023	0.0177 ± 0.0019	0.0177 ± 0.0016	0.0169 ± 0.0013	0.0155 ± 0.0012	0.0138 ± 0.0011	0.0120 ± 0.0011	0.0103 ± 0.0011

Table 5.
Dose rate along & away per unit air-kerma strength ($cGyh^{-1}$ U^{-1}) in liquid water for the BEBIG ^{192}Ir source (model: GI192M11), (+Z toward the source tip and -Z toward the delivery cable).

process. Thus we considered type B uncertainty negligible, and we reduced the uncertainty to statistical uncertainty. A description of the methodology used to estimate the type B uncertainties is mentioned in the updated report of the TG-43 [21] and the HEBD working group report [1].

9. Results and discussion

Table 2 illustrates the results obtained in this study for both the air-kerma strength and the dose rate constant for the two studied sources compared with the quoted results in previous studies. The value calculated for the air-kerma strength for the ^{60}Co source was 3.030 ± 0.002 (10^{-7} U Bq^{-1}), this value agrees well with the value 3.039 ± 0.004 (*10^{-7} U Bq^{-1}) obtained by (Anwarul et al., 2012) in their study [4]. Also, H. Badry et al., (2018) obtained the value 3.042 ± 0.007 (*10^{-7} U Bq^{-1}) in their work [5], and Guerrero et al., (2014) found the value 3.046 ± 0.0070 (*10^{-7} U Bq^{-1}) with a maximum difference of 1.6%. For the ^{192}Ir source [3], the value obtained for the air-kerma strength, which is 1.092 ± 0.004 (*10^{-7} U Bq^{-1}), was compared with the available quoted value from Perez-Calatayud et al., (2012) [1], the difference was within 0.1%.

For the constant of dose rate, we found for the ^{60}Co a value of 1.092 ± 0.001 cGy h^{-1} U^{-1}. This result was compared with the published data quoted in **Table 2**, and we found a maximum difference of 0.5% with Granero et al. (2007) and Anwarul et al., (2012) [2, 4]. The result found for the ^{192}Ir source was 1.109 ± 0.004 cGy h^{-1} U^{-1}, compared with the published data we found a maximum difference of 0.1% with the results obtained by Perez-Calatayud et al., (2012) and Granero et al. (2005) in their work [1, 6].

The radial dose function obtained for the ^{60}Co source in this work (**Table 3**) was in good agreement with the obtained results in other studies using different Monte Carlo codes, especially, for distances (>1 cm) (**Figure 2a**). The **Figure 2b** represented the ratio $g_L(r)_{\text{reference}} / g_L(r)_{\text{this work}}$ calculated to evaluate the deviation of our results from the published data. We observe for the distance greater than 1 cm a maximum relative difference of 0.94% compared with H. Badry et al., (2018) [5]. For the near distance to the source, a maximum relative difference of 6% was found compared with Guerrero et al., (2014) [3], 2.14% compared with H. Badry et al., (2018) [5], and 1.65% compared with Granero et al., (2007) [2]. These results can be attributed partially to the variety of the physics models for radiation transport used in each Monte Carlo code, on the one hand. On the other hand, it can be assigned to the differences in simulated geometries impact. For the ^{192}Ir source (GI192M11), the obtained radial dose function in this work using MCNPX is presented in **Figure 3a**. The comparison with previous works, for the range of distance from 0.25 to 20 cm, was performed using the expression $g_L(r)_{\text{reference}}/g_L(r)_{\text{this work}}$ presented in the **Figure 3b**. For the distance near to the source, we observe a maximum relative difference of 0.40% compared with D. Granero et al., (2005) [6]. For distances greater than 1 cm, the maximum relative difference found was 0.74%. The comparison in the case of the ^{192}Ir was also made with the results obtained for the BEBIG Ir2. A852 source model. The maximum difference was found to be within 1.51% compared with D. Granero et al., (2008) [7] and 0.50% if the comparison is made with Belousov et al., (2014) [8].

The radial dose functions investigated using MCNPX for both ^{60}Co and ^{192}Ir were compared between each other by calculating the ratio $g_L(r)_{\text{Co-60}}/g_L(r)_{\text{Ir-192}}$ illustrated

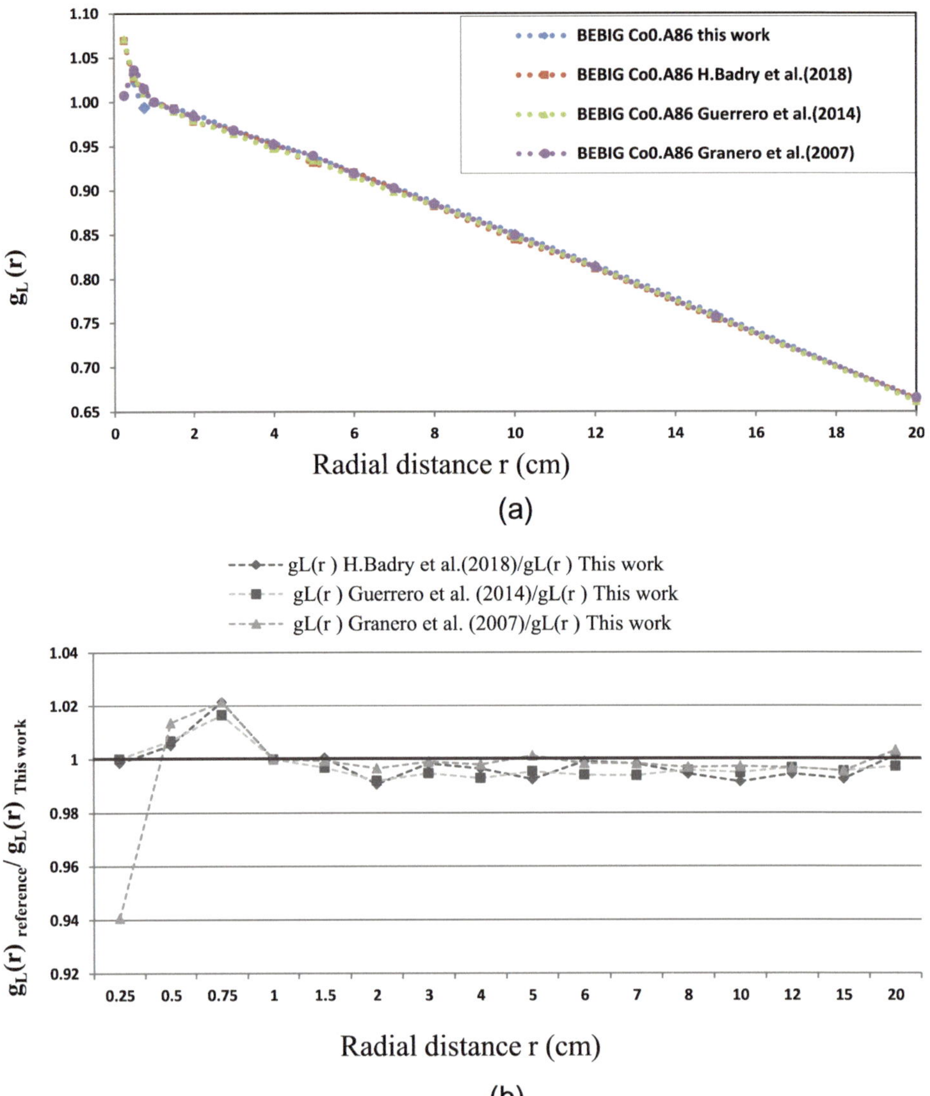

Figure 2.
(a) The curve of radial dose function for ^{60}Co (Co0.A86) obtained with MCNPX compared with the published data for the same model source. (b) The curves of ratio $g_L(r)_{This\ work}/g_L(r)_{reference}$ for ^{60}Co (Co0.A86) obtained with MCNPX compared with previous studies for the same source model.

in **Figure 4**. We observe that the radial dose function calculated for ^{60}Co decrease faster than the ^{192}Ir radial dose function. This difference between the two radial dose functions reached 10% for the distance of 9 cm. This makes the absorbed dose around the two sources different. In addition, regarding the slow decreases of radial dose function for the ^{192}Ir source, we conclude that the ^{192}Ir source could deliver a bit overdoses to the organs at risk more than the ^{60}Co source, especially for tumors of high dimensions in gynecological applications.

The 2D along & away dose rates per unit of air-kerma strength were investigated for the BEBIG ^{60}Co and ^{192}Ir using the same geometry of detectors as mentioned before. The results obtained are tabulated (**Tables 4** and 5), compared with the

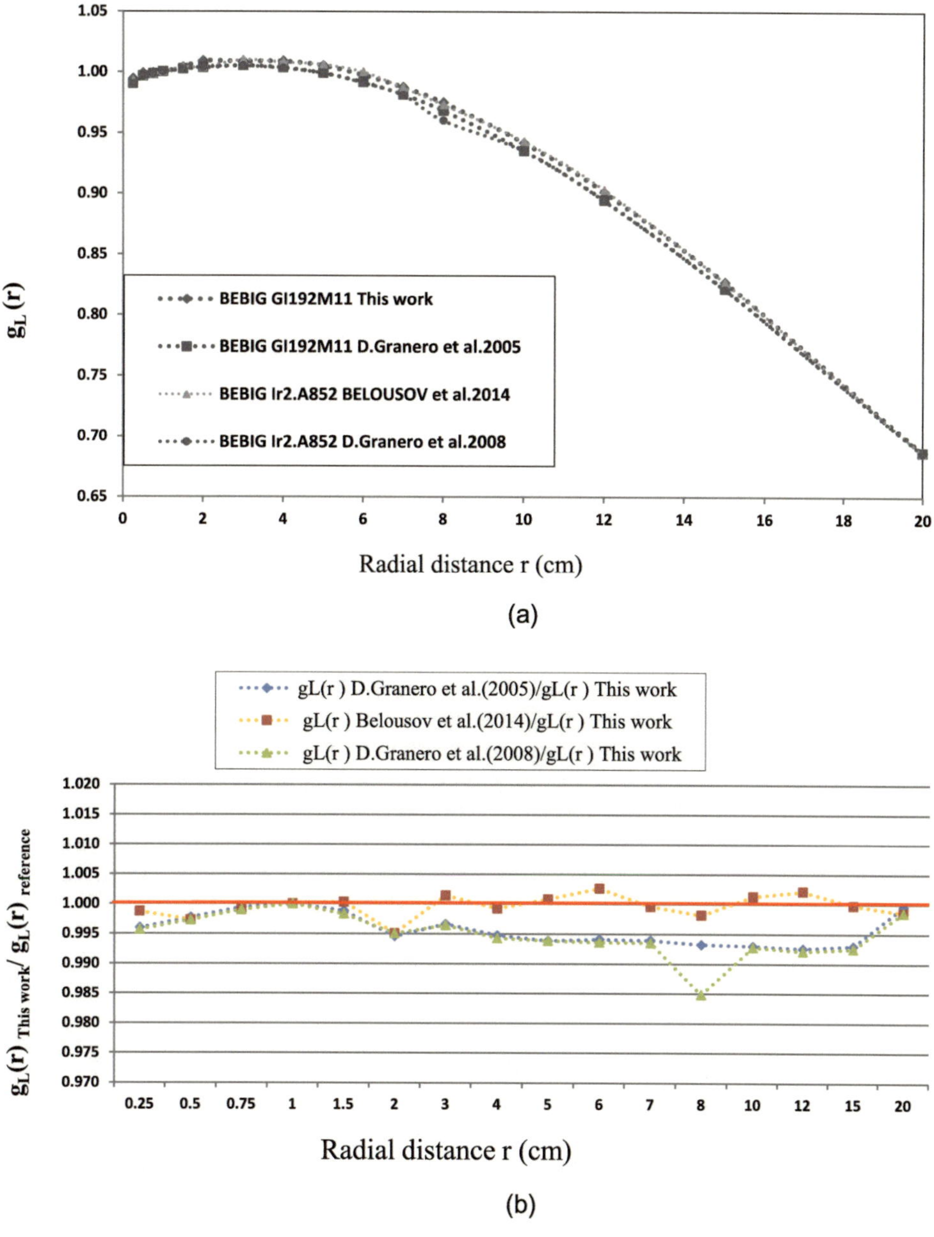

Figure 3.
(a) The curve of radial dose function for ^{192}Ir (GI192M11) obtained with MCNPX compared with the available published data. (b) The curves of ratio $g_L(r)_{\text{This work}}/g_L(r)_{\text{reference}}$ for ^{192}Ir (GI192M11) obtained with MCNPX compared with the previous studies for the same source model, and the source model Ir2.A852.

published data, the results are in good consistency. A comparison was made for the maximum dose rate per U, located in the distance 0.25 cm away in the transversal axis. For the BEBIG ^{60}Co source, we obtain a maximum dose rate per U, which is 16.98 cGy h^{-1} U^{-1} agreeing well with value found by H. Badry et al., (2018) [5], which is 16.55 cGy h^{-1} U^{-1}. Furthermore, Granero et al., (2007) [2] found a value of 15.15 cGy h^{-1} U^{-1}. Otherwise, For the BEBIG ^{192}Ir source, we found a value of 15.41 cGy h^{-1} U^{-1} in this study, and Granero et al., (2005) [6] obtained

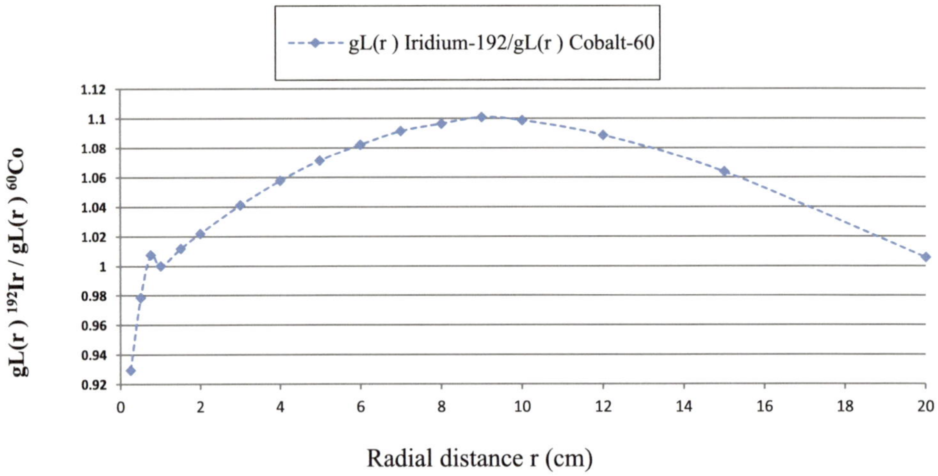

Figure 4.
The curve of the ratio $g_L(r)_{192\text{-}Ir}/g_L(r)_{Co\text{-}60}$*, calculated to compare the differences between the results obtained for* $g_L(r)$ *for both sources Co0.A86 and GI192M11.*

15.50 cGy h^{-1} U^{-1} in their work for the same model. Moreover, Granero et al., (2006) [26] found in their study for Flexisource ^{192}Ir HDR source model a value of 15.56 cGy h^{-1} U^{-1}. Also E. Reys et al., (2016) found 15.57 cGy h^{-1} U^{-1} in their work for the GammaMed HDR Plus ^{192}Ir source model [24].

For the comparison made between the results obtained in this work for both BEBIG sources, we observe that the generated 2D along & away dose rates per unit of air-kerma strength for the near distance to the source are greater for the ^{60}Co than for ^{192}Ir. On the contrary, for distances greater than 1 cm, we observe that the values calculated for the ^{192}Ir source are a little greater than for ^{60}Co source, this difference increases by increasing the distance away in the transversal axis, **Figure 5**. Regarding the contribution of different dosimetric parameters in the treatment planning systems, this difference can be considered negligible within the agreement, concerning the clinic practice for the treated volume [12].

Outside of the treated volume, a study made by Venselaar et al. (1996) mentioned that the absorbed dose in peripheral organs at risk showed opposite behavior (^{192}Ir doses > ^{60}Co doses) at distances near the treated volume in contrast to the behavior (^{192}Ir doses < ^{60}Co doses) at larger distances [13]. In addition, recent study of dose delivered to organs has been calculated on a reference male phantom for a typical implant of the prostate in HDR brachytherapy using Monte Carlo method [14]. For the closest organs, equivalent delivered doses by ^{60}Co were less (8–19%) than for ^{192}Ir. However, increasing the distance beyond 10 cm, high equivalent doses were delivered by ^{60}Co. The overall result is that effective doses per clinical absorbed dose from a ^{192}Ir source are about 18% greater than from a ^{60}Co source [14].

10. Conclusion

In conclusion, the minor differences on the absorbed dose around the two sources observed in the radial dose function decrease and the 2D along & away dose rate per unit of air-kerma strength. For ^{60}Co and ^{192}Ir is considered negligible within the agreement by the specialists evaluated the use of ^{60}Co in the afterloading devices as

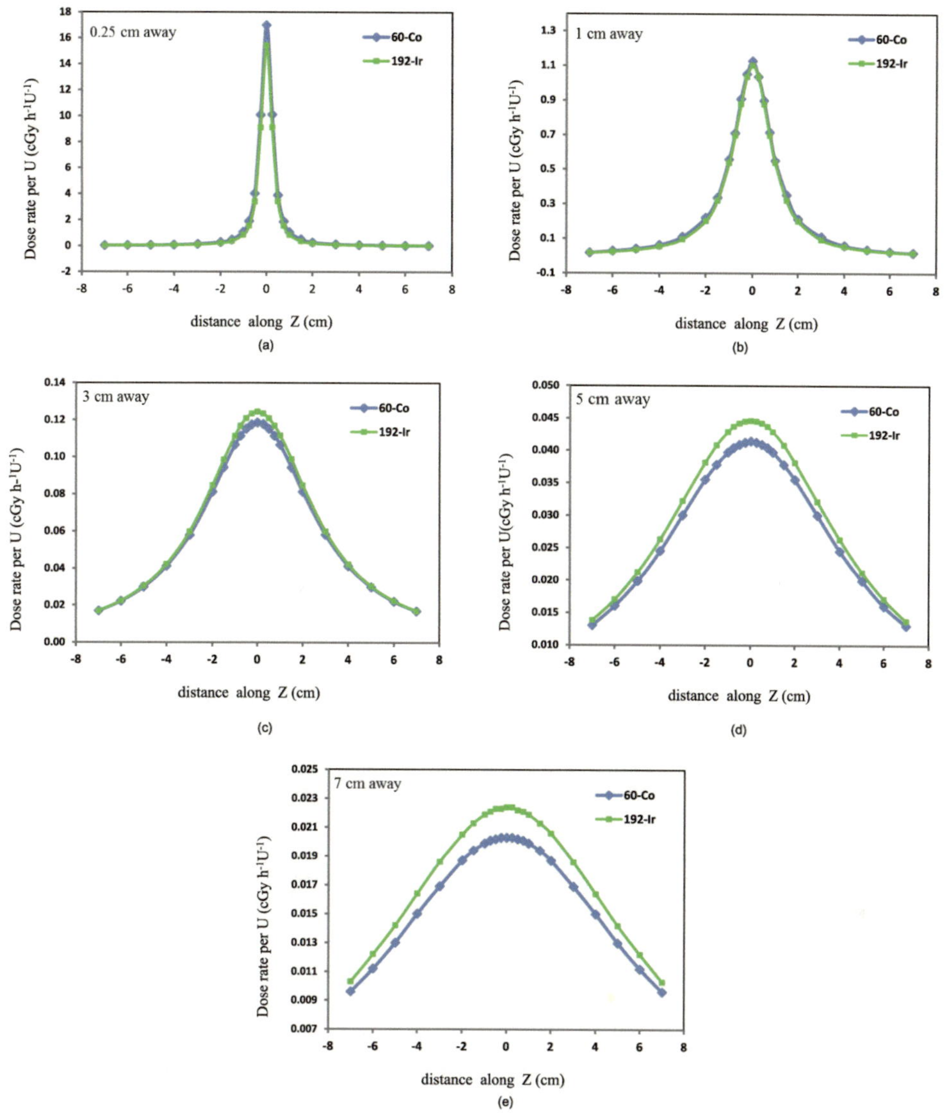

Figure 5.
A comparison between dose rate per unit of air-kerma strength of ^{60}Co (Co0.A86) and ^{192}Ir (GI192M11) sources in different away distances (a = 0.25 cm, b = 1 cm, c = 3 cm, d = 5 cm, e = 7 cm).

^{192}Ir equivalent. Their studies show that there are no significant differences between the two sources concerning the prescribing dose in a typical brachytherapy applications, neither in the treatment planning nor isodose distributions to target on the one hand. On the other hand, economic aspects make the ^{60}Co an important option for clinics over the world. The recent introduction of miniaturized ^{60}Co sources by Eckert & Ziegler BEBIG is considered as a mutation for this nuclide in HDR brachytherapy. The previous study announced that ^{60}Co sources have potential logistical advantages and replacement intervals due to decay. One exchange of the ^{60}Co source required 25 source exchanges for ^{192}Ir, and this reduced operating costs, and makes ^{60}Co a good option to be considered for applications in brachytherapy HDR, especially for the developing countries.

Author details

Said Elboukhari[1*], Khalid Yamni[1], Hmad Ouabi[2], Taoufiq Bouassa[3] and Lahcen Ait Mlouk[4]

1 Faculty of Sciences, EMaMePS, Department of Chemistry, Molay Ismail University of Meknes, Morocco

2 Center of Oncology ALAZHAR, Rabat, Morocco

3 Faculty of Sciences, Mohammed V University of Rabat, Morocco

4 Faculty of Sciences, Molay Ismail University of Meknes, Morocco

*Address all correspondence to: saidhorion@gmail.com

References

[1] Perez-Calatayud J, Ballester F, Das RK, Dewerd LA, Ibbott GS, Meigooni AS, et al. Dose calculation for photon-emitting brachytherapy sources with average energy higher than 50 keV: (HEBD) Working Group report of the AAPM and ESTRO. Medical Physics. 2012;**39**(5):2904-2929. DOI: 10.1118/1.3703892"10.1118/1.3703892

[2] Granero D, Pérez-Calatayud J, Ballester F. Technical note: Dosimetric study of a new Co-60 source used in brachytherapy. Medical Physics. 2007; **34**:3485-3488. DOI: 10.1118/1.2759602

[3] Guerrero R, Almansa JF, Torres J, Lallena AM. Dosimetric characterization of the 60Co BEBIG Co0.A86 high dose rate brachytherapy source using PENELOPE. Physica Medica. 2014;**30**: 960-967. DOI: 10.1016/j.ejmp.2014.06.039

[4] Anwarul IM, Akramuzzaman MM, Zakaria GA. Dosimetric comparison between the micro Selectron HDR 192Ir v2 source and the BEBIG 60Co source for HDR brachytherapy using the EGSnrc Monte Carlo transport code. Journal of Medical Physics. 2012;**37**(4): 219-225. DOI: 10.4103%2F0971-6203.103608

[5] Badry H, Oufni L, Ouabi H, Hirayama HA. Monte Carlo investigation of the dose distribution for 60Co high dose rate brachytherapy source in water and in different media. Applied Radiation and Isotopes. 2018;**136**: 104-110. DOI: 10.1016/j.apradiso.2018.02.016

[6] Granero D, Perez-Calatayud J, Ballester F. Monte Carlo calculation of the TG-43 dosimetric parameters of a new BEBIG Ir-192 HDR source. Radiotherapy and Oncology. 2005;**76**: 79-85. DOI: 10.1016/j.radonc.2005.06.016

[7] Granero D, Pérez-Calatayud J, Ballester F. Monte Carlo study of the dose rate distributions for the Ir2.A85-2 and Ir2.A85-1 Ir-192 afterloading sources. Medical Physics. 2008;**35**: 1280-1287. DOI: 10.1118/1.2868766

[8] Belousov AV, Kalachev AA, Osipov AS. Monte Carlo Calculation of Dosimetry Parameters for a Brachytherapy Source. Biophysics and Medical Physics. 2014;**6**:95-100. DOI: 10.3103/S0027134914060034

[9] Sechopoulos I, Rogers DWO, Bazalova-Carter M, Bolch WE, Heath EC, MF MN-G, et al. RECORDS: Improved Reporting of Monte Carlo Radiation transport Studies: Report of the AAPM Research Committee Task Group 268. Medical Physics. 2017;**45**(1): e1-e5. DOI: 10.1002/mp.12702

[10] Alizadeh M, Ghorbani M, Haghparast A, Zare N, Moghaddas TA. A Monte Carlo study on dose distribution evaluation of Flexisource 192Ir brachytherapy source. Reports of Practical Oncology Radiotherapy. 2015; **20**(3):204-209. DOI: 10.1016%2Fj.rpor.2015.01.006

[11] Elboukhari S, Yamni K, Ouabi H, Bouassa T, Ait-Mlouk L. Technical note: Dosimetric study for the new BEBIG 60Co HDR source used in brachytherapy in water and different media using Monte Carlo N-Particle eXtended code. Applied Radiation and Isotopes. 2020; **159**:109087. DOI: 10.1016/j.apradiso.2020.109087

[12] Andrassy M, Niatsetsky Y, Pérez-Calatayud J. Controversies. 60Co versus 192Ir in HDR brachytherapy: Scientific

and technological comparison. Revista Fis Medicine. 2012;**13**(2):125-130

[13] Venselaar JL, van der Giessen PH, Dries WJ. Measurement and calculation of the dose at large distances from brachytherapy sources: Cs-137, Ir-192 and Co-60. Medical Physics. 1996;**23**: 537-543. DOI: 10.1118/1.597811

[14] Candela C, Perez-Calatayud J, Ballester F, Rivard MJ. Calculated organ doses using Monte Carlo simulations in a reference male phantom undergoing HDR brachytherapy applied to localized prostate carcinoma. Medical Physics. 2013;**40**:033901. DOI: 10.1118/1.4791647

[15] Strohmaier S, Zwierzchowski G. Comparison of 60Co and 192Ir sources in HDR brachytherapy. Journal of Contemporary Brachytherapy. 2011;**4**: 199-208. DOI: 10.5114%2Fjcb.2011.26471

[16] Ballester F, Granero D, Pérez-Calatayud J, Casal E, Puchades V. Monte Carlo dosimetric study of Best Industries and Alpha Omega Ir-192 brachytherapy seeds. Medical Physics. 2004;**31**(12): 3298-3305. DOI: 10.1118/1.1820013

[17] White MC. Further Notes on MCPLIB03/04 and New MCPLIB63/84. Compton Broadening Data For All Versions of MCNP5, LA-UR-12-00018. 2012

[18] Center NND. Nuclear data from NuDat, a webbased database maintained by the National Nuclear Data Center. Upton, NY, USA: Brookhaven National Laboratory; Available from: http://www.nndc.bnl.gov/nudat2

[19] Ballester F, Granero D, Pérez-Calatayud J, Casal E, Agramunt S, Cases R. Monte Carlo dosimetric study of the BEBIG Co-60 HDR source. Physics in Medicine and Biology. 2005;**50**(21):N309-N316. DOI: 10.1088/0031-9155/50/21/N03

[20] International Commission on Radiation Units and Measurements. Tissue Substitutes in Radiation Dosimetry and Measurement. ICRU Report no. 44. Bethesda, MD: International Commission on Radiation Units and Measurements; 1989

[21] Rivard MJ, Coursey BM, DeWerd LA, Hanson WF, Huq MS, Ibbott GS, et al. Update of AAPM Task Group No. 43 Report: A revised AAPM protocol for brachytherapy dose calculation. Medical Physics. 2004;**31**: 633-674. DOI: 10.1118/1.1646040

[22] Borg J, Rogers DW. Monte Carlo Calculations of Photon Spectra in Air from 192Ir Sources. Ottawa, Canada, PIRS-629r: National Research Council; 1999

[23] Williamson JF. Monte Carlo evaluation of kerma at a point for photon transport problems. Medical Physics. 1987;**14**:567-576. DOI: 10.1118/1.596069

[24] Reyes E, Sosa M, Gil-Villegas A, Monzón E. Monte Carlo characterization of the Gamma Med HDRPlus Ir-192 brachytherapy source. Biomedical Physics & Engineering. 2016;**2**:015017. DOI: 10.1088/2057-1976/2/1/015017

[25] Hubbell JH, Seltzer SM. Tables of X-Ray Mass Attenuation Coefficients and Mass Energy-Absorption Coefficients 1 keV to 20 MeV for Elements Z=1 to 92 and 48. Additional Substances of Dosimetric Interest. Gaithersburg, MD: NISTIR; 1995. p. 5632

[26] Granero D, Pérez-Calatayud J, Casal E, Ballester F, Venselaar J. A dosimetric study on the Ir-192 high dose rate Flexisource. Medical Physics. 2006; **33**:4578-4582. DOI: 10.1118/1.2388154

Chapter 7

Intensity Modulated Radiation Therapy Plan (IMRT) Verification Using Indigenous Heterogeneous Phantom

Payal Raina, Rashmi Singh and Mithu Barthakur

Abstract

The dose distribution given by intensity-modulated radiotherapy (IMRT) is highly conformal, compared to conventional radiotherapies; however, due to the presence of the large numbers of fields and irregular shape and size of the treatment segments, the accuracy of IMRT delivery needs to be verified via dose measurement. Different dosimetry techniques are available that measures part of or the whole treatment immediately before a patient is treated and give us the total treatment delivery picture. But the majority of the commercially available phantoms are of homogeneous density, whereas the actual human body is a complex medium of different density patterns. Additionally, the very few heterogeneous phantoms, which are available commercially (i.e., anthropomorphic phantoms) are very costly and are not procured by most of the radiotherapy centers, especially in developing countries. Therefore, an indigenous heterogeneous phantom has been designed to verify the dose distribution prior to patient treatment.

Keywords: heterogeneous phantom, three dimensional conformal therapy, intensity modulated radiation therapy, multi-leaf collimator, Hounsfield unit

1. Introduction

Radiation therapy would not exist without physics. It begins with the discovery of X-rays. This therapy uses ionizing radiation, which is delivered by a linear accelerator. Linear accelerator is a device that uses high-frequency electromagnetic waves to accelerate charged particles, such as electrons to high energies through a linear tube. The high-energy electron beam itself can be used for treating a superficial target, or it can be made to strike a target to produce x-rays for treating a deep-seated target. Radiation therapy works by damaging the DNA of cancerous cells. Photons cause indirect ionization, which happens as a result of the ionization of water, forming free radicals, which then damage. Charged particles, such as electrons, protons, boron, carbon, and neon ions can cause direct damage to target through high-LET (linear energy transfer) [1]. The main focus of physics in radiation therapy is to increase the level of precision and accuracy of dose delivery to the target volume. From the 1950s to the late 1980s, the approach to radiation therapy was based on a two-dimensional (2D) approach. In 2D radiation therapy, plans were created manually, and a single beam used to be given from one to four directions [2]. Shielding blocks were used to collimate the beam. Advances in imaging technology like Ultrasound (US), Computed Tomography (CT), Magnetic Resonance Imaging (MRI), etc. significantly changed the practice of radiation therapy from the 2D method to a Three Dimensional Conformal Therapy (3DCT), which conforms to the high radiation dose with uniform intensity to tumor. For precise shaping of treatment field to the target volume, Multi-Leaf Collimator (MLC) system was developed in place of shielding blocks [3]. Advanced form of radiation therapy called Intensity Modulated Radiation Therapy (IMRT) has been developed in the mid-1990s and early 2000s. IMRT can provide conformal dose distribution compared with 3DCRT [3]. Intensity-Modulated Arc Therapy (IMAT) uses the Multi-Leaf Collimator (MLC) dynamically to shape the fields, as well as rotate the gantry in the arc therapy mode. Intensity-Modulated Arc Therapy (IMAT) was further improved with the addition of variable gantry rotation speeds and dose rates and was introduced as volumetric-modulated arc therapy (VMAT) in 2007. Brief descriptions of all these techniques are discussed in the following sections.

2. Three-dimensional conformal radiotherapy (3DCRT)

Three-Dimensional Conformal Radiation Therapy (3DCRT) means conformal dose distribution in terms of adequate dose to the tumor and minimum possible dose to normal tissue based on 3D anatomic information. The main distinction between treatment planning of 3DCRT and that of conventional radiation therapy is that treatment planning system optimizes dose distribution in accordance with the clinical objectives using anatomic information. Depending on imaging modality, visible tumor, the suspected tumor spread, patient motion uncertainties, critical structures, and relevant landmarks are outlined slice by slice by the radiation oncologist. This technique, however, fails in achieving conformal dose distribution for patient geometries where Organ at Risks (OARs) are located in close proximity to or are even embedded within complicated tumor shapes. Due to lesser conformity of dose distribution in Three-Dimensional Conformal Radiation Therapy (3DCRT), it may be insufficient to allow adequate escalation of tumor dose and there is a need for further improvement. It is possible only with intensity-modulated radiotherapy.

3. Intensity modulated radiotherapy (IMRT)

Intensity-modulated radiotherapy (IMRT) is an advanced form of 3-D conformal radiation therapy that allows modifying the intensity of the beam by considering each radiation beam as multiple rays or beamlets, and assigning different beam strengths to the individual rays [4]. The radiation intensity is adjusted according to the shape, size, and location of the tumor with the use of computer-controlled, moveable "leaves" called Multi-Leaf Collimator (MLC) systems. It consists of pairs of highly absorbing tungsten leaves that can either block or allow the passage of radiation from the many beams as shown in **Figure 1** to deliver a high dose to the target volume and acceptably low dose to the surrounding normal structures [5].

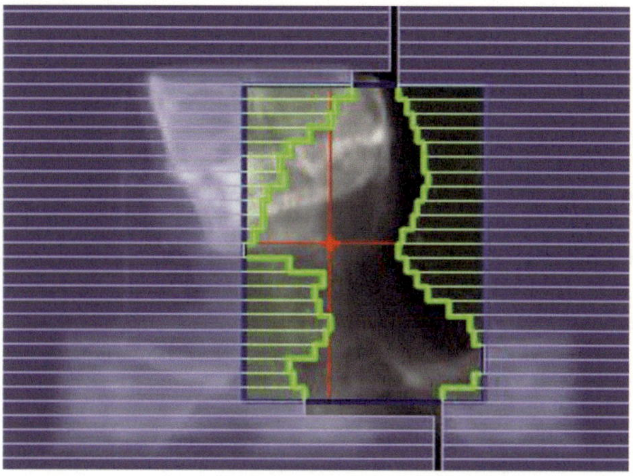

Figure 1.
Multileaf collimators (MLCs).

It uses advanced imaging procedures such as Ultrasound (US), Computed Tomography (CT), Magnetic Resonance Imaging (MRI), Positron Emission Tomography (PET), and PET/CT, to achieve precise treatment modality for cancer patients [6]. A computer-controlled multileaf collimator has been programmed in three different ways to deliver IMRT [7].

1. Multi-segmented Static field Delivery: The patient is treated by multiple fields and each field is subdivided into a set of subfields [8]. The subfields are created by the MLC. The radiation beam is turned off when the leaves are moving from one field segment to another and is turned on only when the leaves reach and stop at the designated segment positions [9]. This method of IMRT delivery is also called "step-and-shoot" or "stop-and-shoot" [10].

2. Dynamic Delivery: In this technique, the beam is kept on while the corresponding leaves sweep simultaneously to produce the desired intensity modulation throughout the treatment delivery [11].

3. Intensity-modulated Arc therapy: Yu has developed an Intensity Modulated Arc Therapy (IMAT) technique. IMAT technique is similar to IMRT, which uses the dynamic mode of dose delivery to shape the fields with gantry rotation and the beam is on all the time [12].

4. Volumetric modulated arc therapy (VMAT)

Intensity-Modulated Arc Therapy (IMAT) has been improved with the addition of variable gantry rotation speeds and dose rates and was introduced as Volumetric-Modulated Arc Therapy (VMAT) in 2007 to describe rotational Intensity Modulated Radiotherapy (IMRT) delivered in a "single arc" [13]. VMAT can provide highly conformal dose distributions and can significantly improve the Intensity Modulated Radiotherapy (IMRT) delivery efficiency. The faster treatments reduce the effects of intra-fractional motion on both tumors and organs, and of course, the shorter treatment times also increase patient comfort. The high plan quality and fast treatment delivery of Volumetric-Modulated Arc Therapy (VMAT) are attractive, and it has been widely applied to many disease sites.

5. IMRT plan verification

In external beam radiation therapy, the energy deposition is three-dimensional in nature. As such particles not only interact with the tumor site but also deposit some of their energy into the adjacent area. Consequently, neighboring normal tissues also receive some amount of radiation dose in this process. Therefore, normal tissue dose tolerance becomes a limiting factor to the success of the treatment. Therefore, a scheduled quality assurance program should be established to verify the plans generated on Treatment Planning System (TPS).

The dose distribution given by Intensity Modulated Radiation Therapy (IMRT) is highly conformal, compared to conventional radiotherapies. But due to the presence of large numbers of fields and irregular shape and size of the treatment segments, the accuracy of Intensity Modulated Radiation Therapy (IMRT) delivery needs to be verified via measurement of dose. Based on the recommendations of the International Atomic Energy Agency (IAEA) published in technical reports series number 277 and 398, there are several techniques to attain accuracy in dosimetry.

Different dosimetry techniques are available to compare the planned dose with delivered dose using an ionization chamber and commercially available phantom, such as slab phantom that measures the point dose at a particular desired reference depth. For reference dosimetry, radiographic film or radiochromic film is placed at a particular depth in slab phantom, and a planned dose is delivered on it. The film quality assurance dosimetry system, for instance, OmniPro IMRT correlates the resultant density of film with the planned dose at each point.

Luminescence dosimetry is also performed using an optically stimulated luminescence (OSL) system and thermoluminescent dosimeters (TLD). It can be also used for in vivo dosimetry in which OSL or TLD are placed on patient's body at reference points for measurement. The electronic portal imaging device is also utilized for reference dosimetry. In addition, many detector-based phantoms are available, for reference dosimetery, such as Accua Check, Delta 4 phantom.

To evaluate an institution's ability to deliver the planned dose to patients, an indigenous heterogeneous phantom has been designed.

6. Designed phantom

The majority of the commercially available phantoms are of homogeneous density, whereas the actual human body is a complex medium of different density patterns [14]. Additionally, the very few heterogeneous phantoms, which are available commercially (e.g. anthropomorphic phantom) are very costly and are not procured by most radiotherapy centers, especially in developing countries. It is known that human body is composed of fat, tissue, bones, and air cavities having different electron densities that influence the interaction of photon and electron energy deposition affecting the dose delivery to a target volume. Therefore, this study was conducted to develop an indigenous heterogeneous pelvic phantom similar to the patient's anatomy and perform a pretreatment verification in a realistic clinical scenario to obtain reproducible dosimetry.

A heterogeneous pelvic phantom was designed, shown in **Figure 2**, which was made of wax, a male pelvic bone (**Figure 3**), water, and borax powder. To construct the phantom, male pelvic bone with a density equivalent to that of human pelvic bone was placed in a cylindrical-shaped container. After placing it, a round plastic ball filled with water was placed for bladder. Borax powder with glue and water was placed below the bladder for rectum. Subsequently, molten wax was poured into it and allowed to solidify. After complete solidification of the wax, the outer container was cut and removed. A cavity was prepared at approximately geometrical center of phantom volume, and a 0.6 cm^3 ion chamber was kept in the same position till the end

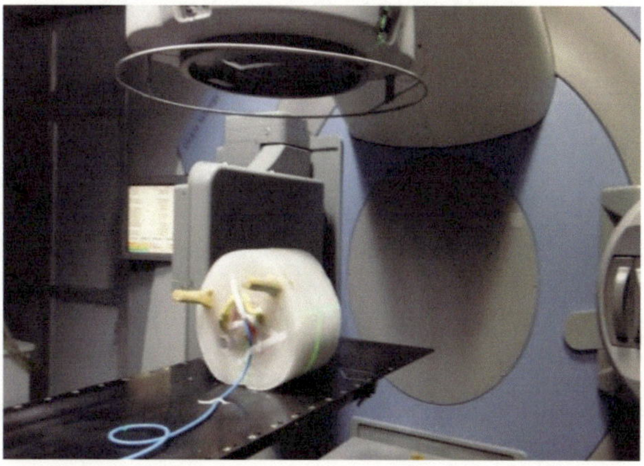

Figure 2.
Designed pelvic phantom.

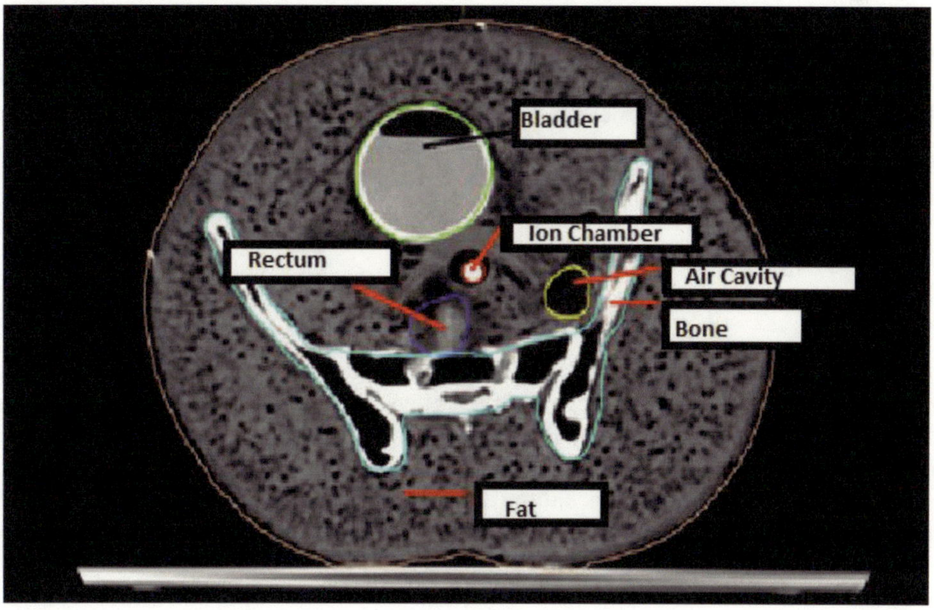

Figure 3.
Male pelvic and femur bone used in developed phantom.

Figure 4.
CT slice of developed phantom with different parts.

of experiment, **Figure 4**. The three fiducially lead markers were put on two bilateral points, and one anterior point was placed on the surface of the phantom in the same cross-sectional plane to make three reference points [15].

Brivo CT 325 2-slice CT (Wipro GE Healthcare, WI, USA) has been utilized for computed tomography (CT) of the phantom and the CT images were taken at a slice thickness of 3 mm for planning purposes. The CT images were imported into the treatment planning system. The width and height were measured using the length measuring tool available in Treatment Planning System (TPS). The mean width and height were measured as 29 cm and 25 cm in CT images of heterogeneous pelvic phantom, respectively. These geometries of the phantom show that it can accommodate delivered beam field sizes and shapes. It allows the establishment of 3D locations. It is easy to transport, set up, align, and takedown in an accurate and efficient manner.

7. Hounsfield unit (CT number)

Computed Tomography (CT) works on the principle of amount of the X-ray energy absorbed. The amount of X-ray energy absorbed is proportional to the density of the body tissue. The computer generates a grayscale image, where the tissue density is indicated by shades of gray. The Hounsfield Unit (HU) is a relative quantitative measurement of radio density used in the interpretation of computed tomography images. The Hounsfield unit was named after Sir Godfrey Hounsfield, recipient of the Nobel Prize, for the invention of Computed Tomography (CT) [16]. It is proportional to the degree of x-ray attenuation and is defined as:

$$HU_{tissue} = \left[\left(\mu_{tissue} - \mu_{water} \right) / \mu_{water} \right] \times 1{,}000 \quad (1)$$

where μ is the linear attenuation coefficient for water and tissue. On the Hounsfield scale, air is represented by a value of −1000 (black on the grayscale) and bone between +700 (cancellous bone) to +3000 (dense bone) (white on the grayscale). The linear attenuation coefficient is a function of both electron density and atomic number of the tissue within a pixel.

8. Electron density

The electron density is the measure of the probability of an electron being present at a specific location. It is calculated from its mass density and its atomic composition.

8.1 Comparison of Hounsfield units and relative electron densities of organs

The Hounsfield Unit (HU) and relative electron density of bone, fat, air cavity, bladder, and rectum in Computed Tomography (CT) images of a heterogeneous phantom and an actual patient were measured and has been given in **Table 1**. All the measurements were calculated by using Computed Tomography (CT) scanner console in terms of mean and stander deviation due to density variation in different CT slices. For the actual patient, a CT image of one patient was taken.

According to the results obtained from the Computed Tomography (CT) images of a heterogeneous pelvic phantom, relative electron densities for bone, fat (wax), air cavity, bladder (water), and rectum (borax powder) were 1.632, 0.896, 0.159, 1.037, and 1.051, respectively. On the other hand, relative electron densities for bone, fat, air cavity, bladder, and rectum were 1.335, 0.955, 0.158, 1.039, and 1.054, respectively, in an actual patient Computed Tomography (CT) image.

8.2 Radiation treatment plan creation

Various Intensity Modulated Radiation Therapy (IMRT) plans for prostrate patients were generated on the Monaco planning system. Plans were created with 5, 7, 9, and 12 coplanar 6MV photon beams. Couch and collimator angles were kept

S.No.	Pelvic Organs	Material	In CT images of a heterogeneous phantom		In CT images of an actual patient	
			HU ± SD	Relative electron density	HU ± SD	Relative electron density
1	Bone	Male Pelvic Bone	1037 ± 179	1.632	556 ± 187	1.335
2	Fat	Wax	−162 ± 45	0.896	−109 ± 108	0.955
3	Air cavity	Air	−846 ± 143	0.159	−847 ± 79	0.158
4	Bladder	Water	−5 ± 5	1.037	−3 ± 8	1.039
5	Rectum	Borax Powder	19 ± 53	1.051	20 ± 26	1.054

CT: Computed Tomography; HU: Hounsfield Units; SD: Standard Deviation.

Table 1.
Comparison of Hounsfield units and relative electron densities of organs.

as 0°for all plans. Calculation parameters such as grid spacing, fluence smoothing, and statistical uncertainty were 0.3 cm, medium, and 1% per plan respectively. Furthermore, the Monte Carlo algorithm was used for the plan optimization, and all the plans were generated in step and shoot mode.

8.3 Gamma analysis

The difference between measured and planned dose distribution is evaluated using quantitative evaluation methods. The Quality Assurance (QA) procedures of Treatment Planning System (TPS) narrated by Van Dyk et al. subdivides the dose distribution comparisons into high and low dose gradients regions, each with a different acceptance standard. In regions of low gradient, planned and measured doses are compared directly, with an acceptance tolerance placed on the difference between the measured and calculated doses. On the other hand, in high dose gradient regions, a small spatial error, either in measurement or calculation, results in a large dose difference between measurement and calculation. Therefore, in the region of high dose gradient, the concept of a Distance-To-Agreement (DTA) distribution is used to determine the acceptability of the dose calculation. The Distance-To-Agreement (DTA) is the distance between a measured data point and the nearest point in the calculated dose distribution exhibiting the same dose. The Dose-Difference (DD) and Distance-To-Agreement (DTA) evaluations complement each other when used as determinants of dose distribution calculation quality.

8.4 Pre-treatment verification

Two kinds of phantoms were chosen for absolute dosimetry of plans already done for the treatment. First one is heterogeneous pelvic phantom developed for radiotherapy quality assurance. Second one was Delta4 phantom (Scandidos, Uppsala, Sweden). CT scan of both the phantoms was done and images were transferred to the Monaco planning system.

After the complete optimization of the Intensity Modulated Radiation Therapy (IMRT), the plans were exported to a pelvic phantom and Delta4 phantom for a pre-treatment verification. After position verification, all Intensity Modulated Radiation Therapy (IMRT) plans were delivered by a linear accelerator.

9. Absorbed dose calculation

There are various methods to achieve accuracy in dosimetry and they are based on International Atomic Energy Agency (IAEA) recommendations published in technical reports series number 277 [17] and 398 [18].

In this study, absorbed dose at reference depth was calculated according to the Technical Reports Series No. 398 (TRS398) of the International Atomic Energy Agency (IAEA) [18] using the relation:

$$D = M_Q \times N_{D,W} \times K_{Q,Qo} \times K_{T,P} \times K_S \times K_{pol} \tag{2}$$

where, M_Q is the electrometer reading (charge), $N_{D,W}$ is the tor, $k_{Q,Qo}$ chamber specific factor, $k_{T,P}$ temperature–pressure correction factor, K_{pol} polarity correction factor, K_S ion recombination factor.

9.1 Chamber calibration factor, $N_{D,W}$

The $N_{D,W}$ is the calibration factor in terms of absorbed dose to water for a dosimeter at a reference beam quality *Qo*. The chamber calibration factor $N_{D,W}$ for the ionization chamber (PTW 0.6 cm^3; TN 30013–006353) is 5.386 x 10^7 Gy/C as obtained by BARC, Mumbai.

9.2 Chamber specific factor, $K_{Q,Qo}$

The $K_{Q,Qo}$is a factor that corrects for the difference between the response of the ion chamber in the reference beam quality Q_o used for calibrating the chamber and in the actual user beam quality Q. The subscript Q_o is omitted when the reference quality is Co-60 gamma radiation i.e. K_Q always corresponds to reference quality Co-60. The chamber-specific factor $K_{Q,Qo}$is 0.99 for the ionization chamber (PTW 0.6 cm^3; TN 30013–006353).

9.3 Temperature pressure correction factor, $K_{T,P}$

The mass of air in the cavity volume is subject to atmospheric variations. The correction factor to be applied to convert the cavity air mass to the reference conditions is given by:

$$K_{T,P} = \frac{(273.2 + T)\ P_O}{(273.2 + T_O)\ P} \quad (3)$$

where, P and T are the cavity air pressure and temperature at the time of the measurements, and P_O and T_O be the reference values (generally 101.3 kPa and 20°C).

9.4 Polarity factor, K_{pol}

The polarity factor is used to correct the response of an ionization chamber for the effect of change in polarity of the polarizing voltage applied to the chamber. It can be accounted for, by using a correction factor

$$K_{pol} = \frac{|M_+| + |M_-|}{2M} \quad (4)$$

where, M_+ and M_- are the electrometer readings obtained at positive and negative polarity, respectively, and M is the electrometer reading obtained with the polarity used routinely (positive or negative).

9.5 Ion recombination factor k_s

The incomplete collection of charge in an ionization chamber cavity owing to the recombination of ions requires the use of correction factor k_s.

$$k_s = a_0 + a_1\left(\frac{M1}{M2}\right) + a_2\left(\frac{M1}{M2}\right)^2 \quad (5)$$

where, a_o = 2.337, a_1 = −3.636, a_2 = 2.299 and M1 and M2 are the electrometer readings at the polarizing voltages V1and V2, respectively, measured using the same irradiation conditions. V1 is the normal operating voltage and V2 is a lower voltage; the ratio V1/V2 is equal to two.

In a pelvic phantom, the dose for each plan was measured by PTW UNIDOS E electrometer connected with 0.6 cm^3ion chambers using Eq. 1 according to International Atomic Energy Agency (IAEA) published, Technical Reports Series-398 (TRS 398) protocol. These measured doses were compared with doses planned on the treatment planning system (TPS).

For Delta4 phantom, TPS calculated dose fluence was compared with measured dose fluence using the gamma evaluation method with critically acceptable criteria of 3 mm Distance-To-Agreement (DTA) and 3% Dose-Difference (DD). Before the evaluation of an Intensity Modulated Radiation Therapy (IMRT) plan, two more measurements were done by delivering 100 cGy with a 10 × 10 cm field at gantry angles of 0° and 90° in order to check the phantom for positional corrections and linear accelerator output constancy.

Table 2 shows the planning parameters, including number of fields, segments, and monitor units, and the percentage variation between planned doses and measured doses for each test case using pelvic phantom.

The gamma analysis results of each test case, including Dose-difference (DD), Distance-To-Agreement (DTA), and Gamma Index passing rates, are presented in **Table 3**.

Plan No.	Algorithm	Energy	No. of fields	Measured Dose	Planned Dose	% Variation
P1	Monte Carlo	6 MV	5	190.34	185.8	2.44 (+)
P2	Monte Carlo	6 MV	7	202.15	207.5	2.58(−)
P3	Monte Carlo	6 MV	9	172.46	176.1	2.07(−)
P4	Monte Carlo	6 MV	12	194.57	191.84	1.42(+)
Mean 2.13						
SD 0.52						

MV: Mega Voltage; SD: Standard deviation.

Table 2.
Percentage variation between planned dose on treatment planning system and measured dose on linear accelerator using heterogeneous pelvic phantom.

Plan No.	Field Number	Segment Number	Monitor Unit	Dose Difference	DTA	Gamma Index
P1	5	14	734.20	79.5%	97.9%	98.4%
P2	7	19	820.31	80.1%	95.1%	97.3%
P3	9	15	775.48	81.4%	94.3%	97.5%
P4	12	14	724.53	80.8%	95.3%	98.8%

DTA: Distance to agreement.

Table 3.
Result of dose difference, distance to agreement and gamma index using Delta4 phantom.

9.6 Test case P1: intensity modulated radiation therapy (IMRT) plan with 5 coplanar beams

The percentage variation between planned dose and measured dose was noted as 2.44% in an indigenously designed heterogeneous pelvic phantom.

Dose distribution at axial projection on the heterogeneous phantom, Delta4 phantom, and on actual patient CT image are shown in **Figure 5a–c** respectively.

The same plan was verified by using Delta4 phantom. The gamma passing rate for test P1 was 98.4%, whereas the pass percentages of Dose-Difference (DD) and Distance-To-Agreement (DTA) were 79.5% and 97.9%, respectively shown in **Figure 5d**.

9.7 Test case P2: intensity modulated radiation therapy (IMRT) plan with 7 coplanar beams

The percentage variation between planned dose and measured dose was noted as 2.58% in the designed pelvic phantom.

The same plan was verified by using Delta4 phantom. The gamma passing rate for test P2 was 97.3%, whereas the pass percentages of Dose-Difference (DD) and Distance-To-Agreement (DTA) were 80.1% and 94.3%, respectively.

9.8 Test case P3: intensity modulated radiation therapy (IMRT) plan with 9 coplanar beams

Similarly, with 9 coplanar beams, the percentage variation between planned dose and measured dose was noted as 2.07%.

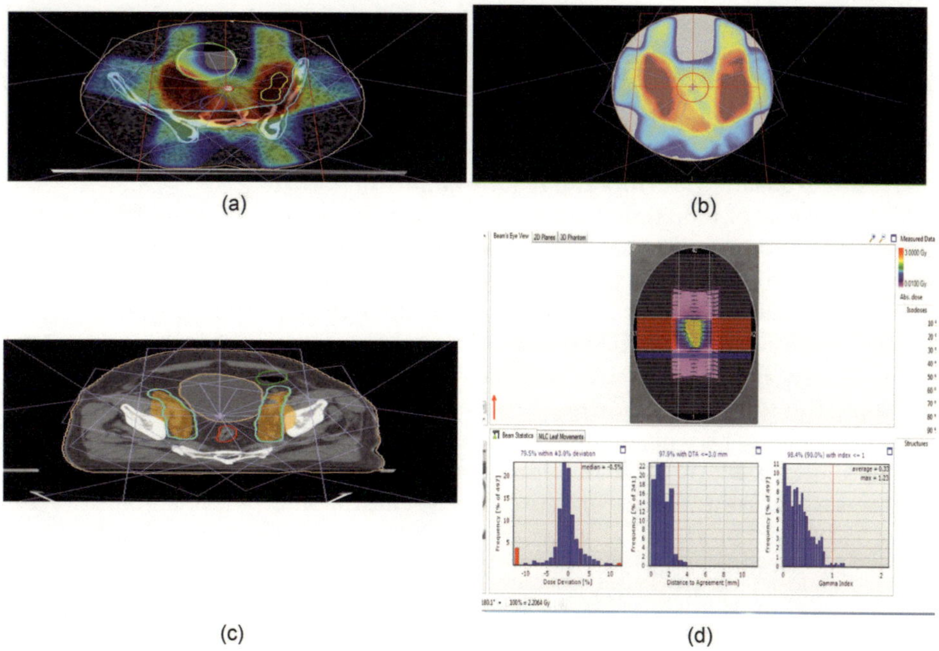

Figure 5.
(a) Dose distribution in heterogeneous phantom, CT slice for test case P1. (b) Dose distribution in Delta4 phantom, CT slice for test case P1. (c) Dose distribution in patient, CT slice for test case P1. (d) Dose distribution, dose deviation, distance to agreement and gamma index of test case P1.

The Dose-Difference (DD) and Distance-To-Agreement (DTA) and Gamma Index were 81.4%, 94.3%, and 97.5% respectively.

9.9 Test case P4: intensity modulated radiation therapy (IMRT) plan with 12 coplanar beams

For the Intensity Modulated Radiation Therapy (IMRT) with 12 coplanar beams, the percentage variation between planned dose and measured dose was noted as 1.42%.

The Dose-Difference (DD) and Distance-To-Agreement (DTA) and Gamma Index were 80.8%, 95.3%, and 98.8%, respectively.

For all the four Intensity Modulated Radiation Therapy (IMRT) plans the percentage variation between the planned dose and measured dose was found to be within the tolerance limit (< ± 3%) prescribed by International Commission on Radiation Units and Measurements (ICRU 83) [19]. Additionally, Gamma evaluation results are based on the critically acceptable criteria of 3 mm DTA and 3% DD given in **Table 3**.

10. Conclusion

In radiation therapy, Quality Assurance (QA) is an essential aspect to ensure that the most accurate treatments are being delivered to a patient. When a new linear accelerator is commissioned at a hospital, the dosimetric parameters, for example, percentage depth dose, radiation beam symmetry, and flatness, are tested to verify the manufacturer's specifications and recommended guidelines. The machine is then periodically tested on a daily, monthly, and yearly basis to make sure that it remains within the specifications. This ensures that the machine continues to deliver what is indicated by a plan. For simple, traditional treatment methods, the Quality Assurance at the machine level is sufficient because possible errors are considered to be acceptable. However, for Intensity-Modulated Radiotherapy and Volumetric Modulated Arc Therapy planned treatments, the increased complexity along with high dose gradients, result in dosimetric errors and require empirical testing.

The main goal of radiation therapy is to deliver a prescribed dose to a target while minimizing the dose to the surrounding normal tissue. As new techniques are developed to achieve this goal, the treatments become more complex and the importance of having accurate dosimetry methods for both initial systems commissioning and ongoing Quality Assurance (QA) increases. Currently, it is mandatory that a patient-specific quality assurance test be performed prior to each new treatment course. Therefore, this study was conducted to develop an indigenous heterogeneous pelvic phantom similar to patient anatomy and perform a pre-treatment verification in a realistic clinical scenario to obtain reproducible dosimetry. Very few heterogeneous phantoms which are available commercially e.g. anthropomorphic phantom are very costly, and are not procured by most radiotherapy centers, especially in low-budget centers in developing countries [20].

In this study, an indigenous heterogeneous phantom was developed using wax for fat, artificial pelvic bone for pelvic bone, water for bladder, and borax powder with glue for rectum. Hounsfield unit and relative electron density of the phantom for different materials used for mimicking the patient were compared with the actual patient pelvic region. A comparison of Hounsfield Unit and electron density shows that the material used for the construction of phantom is almost equal to the

DOI: http://dx.doi.org/10.5772/intechopen.102710

patient tissue heterogeneity as well as shape and tissue content. Materials used for the construction of phantom were locally available, cost-effective, and strong enough to maintain structural integrity.

In this study, Intensity Modulated Radiotherapy was verified using an indigenous heterogeneous pelvic phantom. For validation of heterogeneous phantom, similar plans were also verified using the Delta4 Phantom. The results obtained for all the studies were found to be within the tolerance limit which is <3% as prescribed by the International Commission of Radiation Protection (ICRU 83). This indicates that the phantom can be used successfully for routine patient-specific verification practices.

Author details

Payal Raina[1*], Rashmi Singh[1] and Mithu Barthakur[2]

1 Department of Radiotherapy, Rajendra Institute of Medical Sciences, Ranchi, India

2 Dr. B. Borooah Cancer Institute, Guwahati, India

*Address all correspondence to: payalraina2008@gmail.com

References

[1] Van JD, Smathers JB. The modern technology of radiation oncology: A compendium for medical physicists and radiation oncologists. Medical Physics. 2000;**27**:626-627

[2] Webb S. The physical basis of IMRT and inverse planning. The British Journal of Radiology. 2003;**76**:678-689

[3] Kallman P, Lind B, Eklof A, Brahme A. Shaping of arbitrary dose distributions by dynamic multileafcollimation. Physics in Medicine and Biology. 1988;**33**:1291

[4] Bortfeld T. IMRT: A review and preview. Physics in Medicine and Biology. 2006;**51**:63-79

[5] Purdy A. Dose to normal tissues outside the radiation therapy patient's treated volume: A review of different radiation therapy techniques. Health Physics. 2008;**95**:666-676

[6] Cash CJ. Changing paradigms: Intensity modulated radiation therapy. Seminars in Oncology Nursing. 2006;**22**:242-248

[7] Spirou SV, Chui CS. Generation of arbitrary intensity profiles by dynamic jaws or multileafcollimators. Medical Physics. 1994;**21**:1031-1041

[8] Rehman JU, Ahmad ZN, Khalid M, et al. Intensity modulated radiation therapy: A review of current practice and future outlooks. Journal of Radiation Research and Applied Science. 2018;**11**:361-367

[9] Stein J, Bortfeld T, Dorschel B, Schlegel W. Dynamic X-ray compensation for conformal radiotherapy by means of multileafcollimation. Radiotherapy and Oncology. 1994;**32**:163-173

[10] Brahme A. Optimization of stationary and moving beam radiation therapy techniques. Radiotherapy and Oncology. 1988;**12**:129-140

[11] Boyer AL, Desobry GE, Wells NH. Potential and limitations of in variant kernel conformal therapy. Medical Physics. 1991;**18**:703-712

[12] Yu CX. Intensity modulated arc therapy with dynamic multileaf collimation: An alternative to tomotherapy. Physics in Medicine and Biology. 1995;**40**: 1435-1449

[13] Otto K. Volumetric modulated arc therapy: IMRT in a single gantry arc. Medical Physics. 2008;**35**:310-317

[14] Gurjar OP, Mishra SP, Bhandari V, Pathak P, Patel P, Shrivastav G. Radiation dose verification using real tissue phantom in modern radiotherapy techniques. Journal of Medical Physics. 2014;**39**:44-49

[15] Singh S, Raina P, et al. Dosimetric study of indigenously developed heterogeneous pelvic phantom for radiotherapy quality assurance. Iranian Journal of Medical Physics. 2020;**17**: 120-125

[16] Bushberg JT. The Essential Physics of Medical Imaging. 2nd ed. *Philadelphia, USA*: *Lippincott Williams and Wilkins*; 2002

[17] An International Code of Practice, "Absorbed dose determination in photon and electron beams," IAEA Tech. Series No. 277, Vienna. 1997

[18] An International Code of Practice for Dosimetry based on absorbed dose to water, "Absorbed dose determination in external beam radiotherapy," IAEA Tech. Series No. 398. 2000

[19] International Commission on Radiation Units and Measurements (ICRU), "Prescribing, recording, and reporting photon-beam Intensity-Modulated Radiation Therapy (IMRT). ICRU Report 83," ICRU, Bethesda, MD. 2010

[20] Broerse JJ, Zoetelief J. Dose inhomogeneities for photons and neutrons nearinterfaces. Radiation Protection Dosimetry. 2004;**112**:509-517

Chapter 8

Nuclear Medicine Dosimetry in Paediatric Population

Dimitris A. Verganelakis and Maria Lyra-Georgosopoulou

Abstract

Nowadays, the value of paediatric nuclear diagnostic medical imaging has been well established within the medical community. Despite decades of nuclear medicine practice, studies in nuclear medicine to achieve the lowest possible radiation dose to the patient while ensuring the optimized image quality have to be continued. Numerous studies highlighted a long list of objectives, in order to obtain the minimum possible absorbed dose, achieve short scan times and generate images with a high signal to noise ratio (SNR) and spatial/temporal resolution. For the development of guidelines, it is necessary to study the handling of radiopharmaceuticals, the dose splitting processes, the quality control protocols, the plan design of infrastructures, the availability of optimized dose calibrators for the corresponding radiopharmaceuticals, the development of new more sensitive radiopharmaceuticals, and optimized protocols for diagnostic or therapeutical examination of the patient. Anthropomorphic phantoms are used to model paediatric patients, but anatomical models and their pharmacokinetic data are not applied directly to any specific patient. There is a need for the development of personalized dosimetry in children. Factors regarding age, weight and biological and molecular background of the pathology must be included in paediatric personalized dosimetry. The developmental process of the child, as shape, mass, volume, anatomy, physiological indices (metabolism, heart rate, etc.) and variations due to pathologies should be taken under consideration. Corrections of radiation time of the target organ, in relation to neighbouring tissues, blood supply, estimation of residual activity/time and clearance rate are parameters in the calculations of paediatric dosimetry in nuclear medicine. In hybrid imaging examinations with computed tomography modality, the contribution of absorbed dose from CT to the paediatric patient must also be calculated.

Keywords: paediatric dose phantoms, PDRL, effective dose, image gently, paediatric radiosensitivity, individualize dosimetry

1. Introduction

The aim of this chapter is to briefly describe the topic of paediatric dosimetry in nuclear medicine. Paediatric administrated doses are considered, firstly.

Too high administrated activities increased radiation dose without adding diagnostic information but too low activities may not permit an adequate examination.

The optimal activity amount gives the desired diagnostic information with the minimum patient radiation exposure.

The next session of this chapter refers to the rules and procedures that are established for the evaluation of quantities as absorbed dose, effective dose or Paediatric Dose Reference Levels (PDRL).

In general, patient-specific dosimetry for a child examination in nuclear medicine differs from that of an adult due to different biodistribution and kinetics and variability of body size. It requires the knowledge of many factors like age, weight, biological and molecular background of the pathology and of the developmental process of the child, shape, mass, volume, anatomy, physiological indices.

Different values of instrumental parameters will be used for personalized dosimetric measurements in planar, SPECT, PET, or hybrid (SPECT/CT, PET/CT) examinations in children's studies.

Single Photon Emission Computed Tomography (SPECT) is a nuclear medicine imaging modality that generates 3-dimensional pictures of the examined body.

Positron Emission Tomography (PET) investigates areas of abnormal activity by revealing relative glucose metabolic activity in tissues and organs.

SPECT and PET can be combined with Computed Tomography (CT) or Magnetic Resonance Imaging (MRI) producing the so-called "hybrid" imaging.

Nuclear medicine imaging examinations are performed in paediatric patients for diagnosis of diseases or injuries. The necessity of these examinations by radiopharmaceuticals must be ensured and then should be performed securely.

The International Atomic Energy Agency (IAEA) helps medical professionals to improve quality and safety by providing standards and guidelines, training and information resources. Radiation Protection of Patients (RPOP) is also the leading resource for patients and the public on the safe and effective use of radiation in medicine (IAEA, Radiation Protection of Patients).

World Health Organization (WHO) has edited leaflets and posters titled: 'Nuclear medicine exams in children: what do we need to know?', for patients and families with many simple advices on what do they need to know about nuclear medicine examinations in children.

The performance of a nuclear examination should be based upon the 'justification' radioprotection rule as well as the 'optimization' rule. Both are part of responsible and ethical medical practice.

2. International guidelines in pediatric nuclear medicine (EANM, ICRP, IAEA, image gently, SNMMI, ACR)

Paediatric applications of nuclear medicine provide invaluable information in the diagnosis and follow-up of many pathological disorders as well as in therapy.

In North America, the 'Image Gently' Campaign encouraged the formation of an expert group to overcome the lack of paediatric guidelines and to look into the possibility of developing paediatric harmonized guidelines. These guidelines were approved by the Society of Nuclear Medicine and Molecular Imaging (SNMMI), the Society for Pediatric Radiology (SPR) and the American College of Radiology (ACR).

Although the North American and European Association of Nuclear Medicine (EANM) guidelines use different models, both have concluded in the development of a set of international guidelines, also referred to as "*Pediatric Radiopharmaceutical Administration: Harmonization Guidelines*" [1].

A modified version of the EANM dosage card incorporating the suggested changes is available online. Data on the biokinetics and dosimetry of commonly used radiopharmaceuticals in paediatric nuclear medicine is missing; an appreciable increase, in obtaining more and better data on image quality and biokinetics, focuses on dosimetry as a basis for further improving the recommended administered activities.

Application of the guidelines will allow many paediatric nuclear medicine patients to receive radiopharmaceutical doses lower than those that are traditionally given, resulting in an overall reduction of radiation exposure in these patients.

Paediatric patients are a particular challenge, as body size and the spatial relationships of individual organs can be very different compared to those of a typical adult.

Effective dose is a useful method for assessing the potential radiationally induced effects as a result from various practices within a population group, or more specifically to children within a similar age group, but attention should be paid used when comparing the radiation risks between the groups considering that the organ-weighting factors are averaged over both age and gender [2].

The International Commission on Radiological Protection (ICRP) has published numerous reports addressing the issue of radiation dose with respect to administrated activities in diagnostic nuclear medicine procedures. ICRP Publication 17, published in 1971 and updated in ICRP Publications 53, 80 and 106, address the absorbed dose for various used radiopharmaceuticals. In more details, these reports represent a collection of available data that may be used to estimate radiation dose, expressed as radiation dose to specified target organs, as well as the effective dose, to a population of patients to whom a specific radiopharmaceutical has been administered. They provide conversion factors for the administered activity to effective dose (in mSv/MBq) based on models of patients of different ages, such as 1, 5, 10 and 15 years-old, as well as adults).

Efforts of standardizing and optimising administered activities of radiopharmaceuticals in paediatric nuclear medicine produced fruitful results. Two major guidelines providing administered activities recommended for children, have been developed: one in Europe and the other one in North America.

The European Association of Nuclear Medicine (EANM) issued guidelines for administered activities in children that included a dosage card which provides recommended administered activities for a variety of diagnostic nuclear medicine procedures and radioisotopes correspondingly. EANM's dosage card aim is to secure similar radiation dose levels for all patients undergoing a particular type of nuclear medicine procedure. Therefore, for each radiopharmaceutical, recommended administered activities were calculated (**Figure 1**) so that patients in various age groups receive similar estimated effective doses.

Effective Doses in paediatric PET examinations are included in the following (**Figure 2**), per age group in mSv/MBq.

The Society of Nuclear Medicine and Molecular Imaging (SNMMI) of North America, the American College of Radiology (ACR), the Society of Pediatric Radiology (SPR) and the Image Gently campaign developed also guidelines for dose optimization by identifying best practices [3]. The North American harmonic so-called guidelines are strictly weight-based for 10 out of the 12 procedures included in the guideline, with recommended administered activities corrected for patient size (expressed as mCi/kg or MBq/kg). Consequently, for each nuclear medicine procedure, these guidelines tend to result in similar levels of image noise and thus image quality, for patients of all sizes.

% CW	weight (kg)
0.1	3
0.14	4
0.19	6
0.23	8
0.27	10
0.32	12
0.36	14
0.40	16
0.44	18
0.46	20

$$\text{Activity (MBq)} : \frac{\text{Adult D[MBq]} \times \text{CW[kg]}}{70}$$

Effective Doses in Paediatric NM exams (mSv)

Radio-pharmaceutical	newborn	1	5	10	15
	years				
Tc99m-DMSA	1.29	0.55	0.53	0.68	0.85
Tc99m-DTPA	0.6	0.38	0.54	0.66	1.36
Tc99m-MAG3	0.4	0.18	0.21	0.27	0.66
Tc99m-MDP	2.56	1.84	1.79	2.06	2.22
I123-MIBG	11.76	5.23	3.93	4.7	6.6

Figure 1.
EANM dosage card- Radiopharmaceutical activities are calculated for administration to paediatric patients by weight coefficient in all age groups. Effective doses are estimated for various diagnostic examinations.

Effective Doses in paediatric PET examinations	age / years 15	10	5
	mSv/MBq	mSv/MBq	mSv/MBq
Methyl-C-11 Thymidine	0,0044	0,0068	0,011
2-C-11 Thymidine	0,0034	0,0053	0,008
O-15 water	0,0014	0,0023	0,004
F-18-FDG	0,025	0,036	0,050
Ga-67-Citrate	0,130	0,200	0,330
Rb-82	0,007	0,010	0,018
Rb-81	0,037	0,057	0,110

Figure 2.
Effective doses in paediatric PET examinations by common PET radiopharmaceuticals for age groups 5-, 10-, and 15- (ICRP publications).

European and North American guidelines differ due to the different models used to develop them. In addition, for some radiopharmaceuticals, there are considerable variances between the two guidelines in the reference adult administered activities, which are used in order to calculate children's activities.

Differences in the effective doses resulting between the two 'schools' were more pronounced in younger patients. For ages 1 year or 5-year olds, the EANM's administered activities result in an estimated effective dose at least 20% greater than that provided by the North American guidelines.

The critical organ is independent of the administered activity of radiopharmaceuticals. The most common critical organ is the urinary bladder. At the administered activities recommended by the two guidelines, the highest radiation absorbed doses to other critical organs are those produced by Tc99m-MDP to bone and I123-MIBG to the liver. Normal renal function is assumed when dose estimates are calculated.

Age-specific or disease-specific alteration in organ function can change the biokinetics of a radiopharmaceutical and thus change radiation exposure.

The ICRP allows an adjustment for abnormal renal function or for unilateral ureteral blockage when calculating the absorbed radiation dose from renal imaging agents. For example, infants with biliary atresia have an underdeveloped or absent gallbladder, so the gallbladder is unlikely to be the critical organ during a performance of hepatobiliary scintigraphy in these children.

2.1 Image gently

The Image Gently Alliance was formed to help change practice and increase awareness about radiation exposure to children by medical imaging. The effort of Image Gently Alliance was supported by SNMMI, the SPR and the ACR.

A Nuclear Medicine Working Group has assisted to *standardize* radiopharmaceutical administered activities in the practice of paediatric nuclear medicine across North America and to *harmonize* these practices with those in Europe.

The Nuclear Medicine Global Initiative project (NMGI) was formed in 2012 by 13 international organizations to promote human health by advancing the field of nuclear medicine and molecular imaging. The first project focused on the standardization of administered activities in paediatric nuclear medicine and resulted in two articles [4, 5].

Guidelines have a positive effect on the practice of many nuclear medicine departments dealing with children. Resources useful for radiation dose estimation of paediatric nuclear medicine examinations can be obtained in *Paediatric Injected Activity Tool* (SNMMI) for estimation of injected activity in children and *Nuclear Medicine Radiation Dose Tool* (SNMMI) for an approximate effective dose estimation either by ICRP185, 2015 model or by RADAR model 2017 in various paediatric nuclear medicine examinations.

3. Paediatric dose phantoms

Paediatric model-derived pharmacokinetics to compare absorbed dose and effective dose estimates for F18–FDG in paediatric patients, using S values generated from two different geometries of computational phantoms; Cristy-Eckerman stylized phantoms (C–E) and University of Florida/National Cancer Institute (UF/NCI) hybrid computational phantoms.

Time–integrated activity coefficients of F18–FDG in brain, lungs, heart wall, kidneys and liver, retrospectively, were calculated. The absorbed dose calculation was performed in accordance with the Medical Internal Radiation Dose (MIRD) method using S values generated from the UF/NCI hybrid phantoms. The effective dose was computed using tissue–weighting factors from ICRP publication 60 and 103 for the C–E and UF/NCI, respectively.

Differences in anatomical modelling features among computational phantoms used to perform Monte Carlo–based photon and electron transport simulations for F18, effect internal organ dosimetry computations for paediatric nuclear medicine studies.

Paediatric pharmacokinetic data are collected for diagnostic imaging agents, relevant to paediatric studies and the field conversions from older stylized phantoms to more detailed computation hybrid phantoms were created. The effective doses, computed by the UF/NCI hybrid phantom S values, were different than those seen using the C–E stylized phantoms for newborns, 1-year-old and 5 years old, **Figures 3** and **4** [6].

Since hermaphrodite *Cristy-Eckerman* phantoms are used to represent the newborn, 1-year-old and 5-year-old anatomies, the OLINDA/EXM (*Organ Level INternal Dose Assessment/EXponential Modelling*) code developed by the Radiation Dose Assessment Resource (RADAR) Task Group of the Society of Nuclear Medicine, output for these age groups provides organ–absorbed doses for both paediatric male and female tissues.

In contrast, the University of Florida hybrid phantoms are gender-specific and these tissues are specifically modelled age-wise. The dose estimates for breast and ovaries obtained by the University of Florida F/NCI hybrid phantom were higher for newborn, 1-year-old and 5-year-old ages. The effective dose coefficient computed by OLINDA/EXM version 1.0 uses an effective dose coefficient that is based on radiation and tissue weighting factors specified in ICRP Publication 60 (1991). Later publication 103 (2007), readjusted the tissue weighting factor for breast from 0.05 to 0.12 and for gonads from 0.20 to 0.08.

The understanding of transitioning from the older phantoms and tissue–weighting factors to the most recently updated phantoms that are now being adopted by ICRP is essential (OLINDA/EXM version 2.0).

The OLINDA/EXM has standardized dose calculations for diagnostic and therapeutic radiopharmaceuticals. The previous generation of anthropomorphic phantoms based on the Oak Ridge models, employed geometrical shapes in order to define the body and its organs.

Nowadays, these models have been replaced with realistic, Non-Uniform Rational B-Spline (NURBS) type models based on the recent standardized masses defined by the ICRP in its Publication 89. NURBS is a mathematical model using B-splines that is commonly used in computer graphics for representing curves and surfaces.

These and other new models have been implemented in a new version of the OLINDA/EXM 2 code. The new generation of models is now available in the

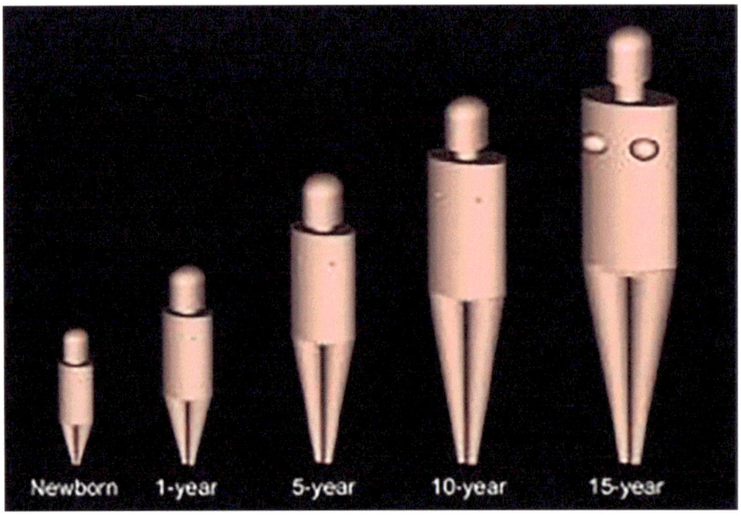

Figure 3.
Three-D visualization of Cristy-Eckerman (C-E) stylized phantoms [6].

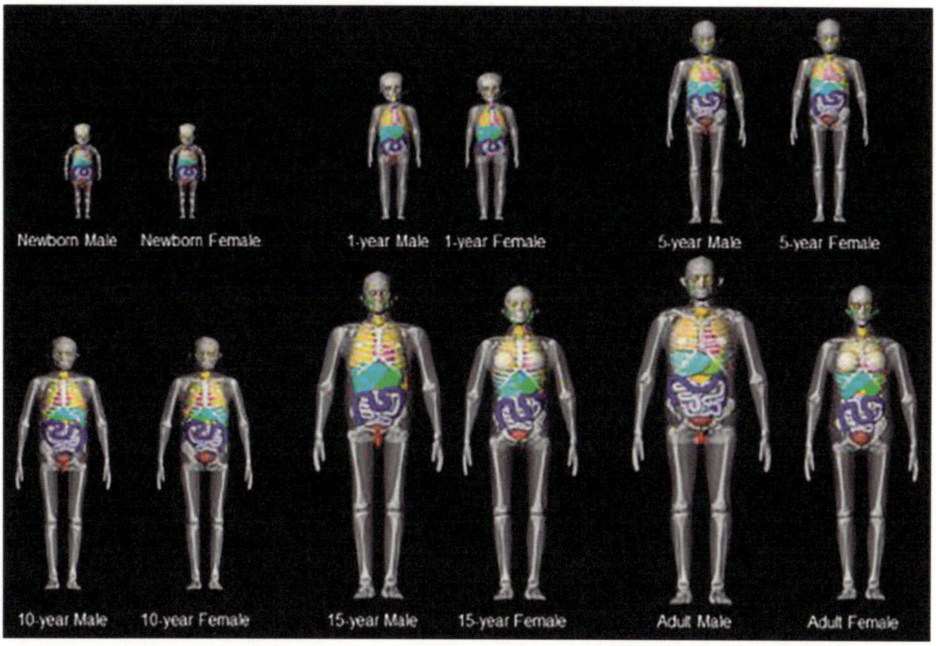

Figure 4.
Three-D visualization of University of Florida/National Cancer Institute (UF/NCI) hybrid computational phantoms for various age groups [6].

OLINDA/EXM code and represents a significant improvement in standardized dose calculations. OLINDA/EXM version 2.0 employs realistic NURBS-style phantoms [7].

ICRP in Publication 143, Paediatric Computational Reference Phantoms, 2020, has adopted a set of reference phantoms that were derived from the University of Florida phantoms. ICRP phantoms will be used to calculate ICRP dose coefficients. The publication is supported by a series of annexes. The last annex gives a description of the electronic files available for download and use of each of the 10 paediatric reference computational phantoms.

A reference set of phantoms and dose coefficients for external exposures and intakes of radionuclides will promote consistency in the assessment of doses.

4. Paediatric dose estimations

4.1 RADAR—OLINDA/EXM 2

Based on the OLINDA/EXM version 2.0 software and on 2007 recommendations of the ICRP, a new generation of voxel-based, realistic human computational phantoms was developed by the RADAR committee of the SNMMI.

It was used to develop the dose estimates as well as the most recent biokinetic models. These estimates will be made available in electronic form and can be modified and updated, as models are changed and new radiopharmaceuticals are added, MIRD Pamphlet No. 21 [8].

RADAR Dose Estimates Report in 2018 based on OLINDA/EXM Version 2.0 for Radiopharmaceutical Dose Estimates [9].

- The MIRD method uses the term A˜ (cumulated activity) for the time–activity integral and presents the dose factor by the S factor.

- RADAR uses the terms N (number of disintegrations) and dose factor, respectively.

The ICRP has a method for internal dose calculations, originally described in ICRP Publication 30. This schema has been repeated, with modifications, several times, the latest being in ICRP Publication 130.

In many ICRP documents, slightly different names are given to some terms, but all the concepts are identical [8, 9].

OLINDA/EXM 2.0 used biokinetic models for 100 radioisotopes and adult and paediatric phantoms in order to develop dose estimate tables. Data within the ICRP task group on radiopharmaceutical dosimetry was considered. Tables for males and females were generated for 1-y olds, 5-y olds, 10-y olds, 15-y olds and adults.

The dose estimate tables give male and female dose values for approximately 25 target organs, as well as sex-averaged values for the five phantom-ages considered (1-y olds, 5-y olds, 10-y olds, 15-y old and adults). In these estimations, individual organ doses are given in units of equivalent dose (e.g., mSv) and not of absorbed dose (e.g., mGy), as quality factors applied may be non-unity for some emitters. For example, OLINDA/EXM 2.0 uses:

- for emissions a default radiation weighting factor.

- variable radiation weighting factors adjusted by the user.

- effective doses that are expressed in the same units as equivalent doses, by applying individual tissue weighting factors.

- a bone model that is the same as that used in OLINDA/EXM versions 1.0 and 1.1.

- unlike the Cristy-Eckerman phantoms, no breast tissue that was assigned in children 10 years old or younger.

- several new organs that have been defined in the RADAR phantoms.

Biokinetic models for nearly 100 radiopharmaceuticals can be used with the OLINDA/EXM 2.0 paediatric phantoms to develop dose estimate tables. Male and female tables for 1-year-olds, 5-year-olds, 10-year-olds, 15-year-olds can be generated.

In ICRP Publication 103, a sex-averaged rule is described for the development of relative data.

Individual organ doses are given in units of equivalent dose (mSv) and not in units of absorbed dose (mGy).

5. Paediatric diagnostic reference levels (PDRLs)

The International Commission on Radiological Protection (ICRP) Publication 73 was first to introduce the term 'Diagnostic Reference Level' (DRL) in 1996, a concept that was further developed further. The DRL has been proven to be an effective tool towards the optimisation of protection in the medical exposure of patients in diagnostic procedures.

5.1 Planar-SPECT imaging procedures

For planar nuclear medicine imaging, DRLs have been set either by administered activity (MBq) or by administered activity per body weight (MBq/kg).

For SPECT imaging procedures, DRL values should be used in the same way as for planar nuclear medicine procedures. DRL values for SPECT studies are usually slightly higher than for the same radiopharmaceuticals used for planar imaging.

5.2 Positron emission tomography (PET)

Specific radiopharmaceuticals are used for PET imaging, depending on the scope of the study. F18-fluorodeoxyglucose (F18-FDG) is used for diagnosing, staging and assessing therapeutical schemes in cancer, inflammation, viable myocardium and brain diseases by revealing relative glucose metabolic activity in tissues and organs. N13-ammonia or Rb82-chloride assesses myocardial perfusion. Ga68-DOTATATE and DOTATOC in neuroendocrine tumours reflecting the status of somatostatin receptors. As the physical half-lives of radionuclides and biological half-times of radiopharmaceuticals are different, DRL values have to be set for each one.

European guidelines provide a calculation system according to body weight, image acquisition method (two-, or three-dimensional), scan speed (minutes per table position) and table overlap during the following PET acquisitions.

5.3 Hybrid imaging (PET-CT, SPECT-CT and PET-MRI, SPECT-MRI)

PET and SPECT have been combined with the modality of CT generating the so-called hybrid systems of PET-CT and SPECT-CT accordingly. Nowadays, they have been combined with the Magnetic Resonance Imaging (MRI) modality too, as these combinations increase diagnostic accuracy by providing both functional and anatomical images of the body.

The acquisition of accurately co-registered anatomical and functional images is a major strength of combined modality (hybrid imaging) devices. The patient dose from a PET-CT or SPECT-CT examination is the combination of the radiation exposures caused by the radiopharmaceutical and by the CT study via the exposure to ionising radiation.

The MRI component of PET/MRI or SPECT/MRI does not increase the patient dose considering that it uses non-ionising radiation, so from a radiation protection point of view, this hybrid imaging is preferable in paediatric examinations.

In the framework of a PET/CT or SPECT/CT, the CT portion of the examination consists of a localiser radiograph and the helical CT scan. If a CT is solely performed for attenuation correction and co-localisation, the acquisition parameters (tube current, voltage, slice thickness, rotation time, and pitch) should be selected in order to minimise the patient's radiation exposure. A low-dose CT used in hybrid imaging is sufficient for attenuation correction and anatomic localisation and proper for paediatric examinations.

5.4 Paediatric diagnostic reference level of examinations in nuclear medicine (PDRL)

Establishing Dose Reference Level values for children is more challenging than for adults, due to the broad range of sizes of paediatric patients. Weight in children can vary by a factor of more than 100 from a premature infant to an obese adolescent. The amounts of radiation used for examinations of children can vary extremely due to the great difference in children's size and weight.

Patient age groups have been used in the past in order to establish Paediatric DRL values. However, it has been recognized that age alone is not a representative parameter. Weight categories have to be included and should be used whenever possible. The difference in patient dose due to patient weight is expected and therefore weight ranges are recommended for establishing Paediatric DRL values.

Age groups around the ages of 0, 1, 5, 10 and 15 years can be used if age is the only available quantity. For examinations including the head, age grouping is recommended for establishing PDRL values. In paediatric imaging, sufficient data is an issue and therefore it has been suggested that the DRL quantity could be a function of patient's weight.

For nuclear medicine imaging, the DRL quantities and DRL values are set as administered activity per body weight (MBq/kg) as a practical and simple approach.

Activities for administration should be adjusted based on size or weight associated factors.

When regional or national DRL values, relevant for paediatrics, are not available, the local practice may be compared with appropriate available published data.

For CT used in a hybrid system SPECT/CT or PET/CT, the DRL quantities are Computed Tomography Dose Index (CTDIvol) and Dose Length Product (DLP), based on calibration with a 32-cm-diameter phantom for body examinations and a 16-cm-diameter phantom for head examinations.

The CTDIvol and the DLP are common methods to estimate a patient's radiation exposure from a CT procedure. The exposures are the same regardless of patient size, but the size of the patients is a factor in the overall patient's absorbed dose.

The unit of CTDIvol is the gray (Gy) and it can be used in conjunction with patient size to estimate the absorbed dose. The CTDIvol and absorbed dose may differ by more than a factor of two for small patients such as children. On the other hand, DLP measured in mGy.cm is a measure of CT tube's radiation output/exposure. It is related to volume CT Dose Index (CTDIvol). $CTDI_{vol}$ represents the dose through a slice of an appropriate phantom and DLP accounts for the length of radiation output along the long axis of the patient. DLP = ($CTDI_{vol}$) [*in units: mGy.cm*]. DLP does not take into account the size of the patient and is not a measure of absorbed dose or the patient's effective dose.

The effective dose depends on factors including patient size and the region of the body being scanned. Values for these quantities should be obtained from patient examinations. Most CT scanners permit the determination of effective diameter or patient equivalent thickness. This is an additional improvement for setting Paediatric DRL values.

Size Specific Dose Estimate (SSDE) measured in mGy, is a method of estimating CT radiation dose that takes a patient's size into account. SSDE may be used in addition to the recommended DRL quantities as an extra source of information for the evaluation of the absorbed dose value.

Results from the largest international dose survey in paediatric computed tomography (CT) in 32 countries are included in ICRP Publication 135 where international DRL for Paediatric computed tomography were established [10]. Patient data were recorded among four age groups: <1 year, 1–5 years, <5–10 years and <10–15 years.

5.5 Views related to paediatric DRLs

The risk of harmful radiation effects is greater in children than in adults and optimisation of paediatric imaging is of particular importance because they have a longer life expectancy during which these effects may appear.

The amount of radiation used for examinations of children can vary greatly due to the excessive difference in patient size and weight from neonates to adult-sized adolescents.

Variation in patient radiation dose for two paediatric patients with the same size, same exposed area of anatomy should be the minimum. If not, this could be due to poor technique, or failure to adapt imaging protocols to account for both paediatric diseases and paediatric patient sizes. Weight or size-adjusted paediatric DRL values are therefore particularly important in optimization.

A number of factors need to be considered when communicating the development of DRL values for children. Some parameters are the same for adults and children. These include the choice of DRL quantities, the percentile of the distribution of the DRL quantity and whether to collect data from patient examinations or from measurements with phantoms.

DRL values for children, there cannot be as a single standard patient due to the large size range of paediatric patients [11].

Weight in children can vary by a factor of more than 100, from that of a premature infant.

Within the first 6 months of life, a typical baby's body weight doubles, and during the first year, it increases 3-fold. Ideally, five or more size ranges should be established between premature to infants (newborn, >1, >5, >10 and >15 years) [12].

It is preferable creation of groups based on paediatric patient body size and that body size be determined for individual patients before performing diagnostic imaging procedures by radiation sources.

In 1999, the European Commission issued Radiation Protection 109 (RP 109) with the title: 'Guidance on diagnostic reference levels (DRLs) for medical exposure'. This document indicates the critical need of establishing DRLs for high-dose medical examinations of patients sensitive to radiation, such as children. This work used average-sized adult phantom or standard size phantoms.

However, the same approach has not been considered appropriate for children due to the wide variation in body habitus.

DRL values for paediatric patients are only available for some common radiological examinations and there is a need to generate appropriately more.

The European Commission recognized this need and approved the 27-month tender project, European Diagnostic Reference Levels for Paediatric Imaging (PDRL) on the establishment of European DRLs for paediatric patients in December 2013. PDRL is coordinated by the European Society of Radiology (ESR, *Eurosafe Imaging)*, **Figure 5** [12].

The Japanese Society of Nuclear Medicine (JSNM) in 2014 has published the consensus guidelines for paediatric nuclear medicine. JSNM proposes dose optimization in paediatric nuclear medicine studies and widely discusses imaging techniques for the appropriate conduct of paediatric nuclear medicine procedures, considering the features of children imaging in order to produce harmonic PDRL [13].

Scientists in nuclear medicine departments must be familiar with

- the increased radiosensitivity of children,

- the risks of low dose radiation,

Paediatric Diagnostic Reference Levels(MBq)

Radiopharmaceutical	newborn	1 year old	5 years old	10 years old	15 years old
Tc99m-DMSA	15	15	25	46	76
Tc99m-DTPA	20	28	50	91	152
Tc99m-MAG3	15	15	17	32	53
Tc99m-MDP	41	70	126	228	378
I123-MIBG	80	80	101	182	303

Figure 5.
The calculated PDRLs may help in the standardization of the appropriate activity in paediatric nuclear medicine [18].

- the patterns of dedicated clinical results when radiation activities in paediatric patients are minimized.

Regarding the reduction of radiation exposure to paediatric patients, continuous education and thoughtful application of techniques for radiation dose management may lead to the improvement of risk-benefit ratios when performing diagnostic imaging in children by radiopharmaceuticals.

Technology provides options such as new software and new hardware (collimators, computer components, etc.) for reducing radiation exposure while maintaining image quality driving to a minimum variation in PDRL values, globally.

6. Radiobiology in nuclear medicine and molecular imaging

Paediatric patients are referred to nuclear medicine from nearly all paediatric specialities including urology, oncology, cardiology, gastroenterology or orthopaedics. Radiation exposure is associated with a potential small risk of inducing cancer in the patient, later in life; this danger is higher in younger patients.

In the field of *nuclear medicine imaging*, which uses very small amounts of radioactive substances *(radiopharmaceuticals)* to diagnose and treat disease, the radiopharmaceuticals injected into the patient's body are detected in very precise images of the part of the body being imaged.

6.1 The role of radiobiology in nuclear medicine

In 2021, the EANM published a position paper on the role of radiobiology in nuclear medicine [14]. For that paper, a group of EANM radiobiology, physics and dosimetry experts summarized the main issues concerning radiobiology in nuclear medicine. The position of the EANM is that radiobiology will contribute to the *optimization* of radiotherapy to ensure that they are effective and safe for each individual patient, considering age and weight.

There is a need to generate and apply more radiobiologic knowledge specific to nuclear medicine diagnostic and therapeutic procedures, as DNA damage induction and repair strongly because of the comparatively low dose rates varying over time with physical decay and kinetic clearance.

While the role of radiobiology for diagnostics remains to be clarified, its role in the benefits of radiopharmaceuticals in therapy is clear.

It is expected that a better understanding of radiobiological parameters can contribute to fully exploiting the abilities of new and existing nuclear medicine applications; how can be effective and safe for each individual patient, child or adult. Radiobiology plays an important role in supporting *optimizations*, in an increase of the use of radiopharmaceuticals for diagnostic or therapeutic nuclear medicine.

A better understanding of radiobiologic parameters will enhance the capabilities of new and existing nuclear medicine applications in adults and paediatric patients. There is a need to better define the dose-effect relationships of radiopharmaceutical radiation in tumours and normal tissue. To reach this target, the EANM recommends a strong link between all scientists involved (Radiobiologists, radiochemists, radiopharmacists, medical physicists, and physicians). So, an improved understanding of the biological processes, with special regard to the effects of ionizing radiation to normal tissues and tumours, for any living matter, will be gained.

When ionizing radiation interrupts living matter, it deposits energy along its path leading to atomic ionization, thereby damaging biological molecular structures (**Figure 6**).

DNA damage induced by radiation is considered critical. DNA, as well as proteins, lipids and metabolites can potentially be modified by ionizing radiation. As first action, absorption of ionizing radiation will occur at the site of the atoms of the cellular molecules. Following ionization events may cause the breakage of chemical bonds. It may also convert atoms and molecules into free radicals with very sensitive unpaired electrons that can further interact with close molecules, after which a damaging sequence may occur.

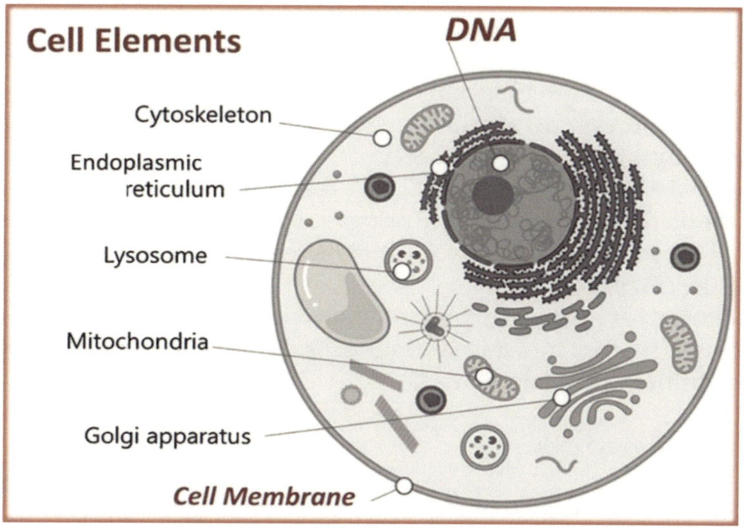

Figure 6.
Interaction of ionizing radiation with cellular matter- DNA and others. DNA and other cell elements are potential targets for ionizing radiation damage. Ionizing radiation also influences cell signalling pathways like oxidative stress, cell death and survival pathways, premature ageing and inflammation [14].

6.2 Molecular imaging: how it works

Molecular imaging provides detailed images at the molecular and cellular levels. Molecular imaging indicates how the body is functioning and gives the prospect to measure its chemical and biological processes. It offers exclusive insights into the human body that patients can obtain *personalized* care. In diagnostic molecular imaging, diseases are identified in the earliest stages and the exact location can be determined, avoiding more invasive procedures such as biopsy or surgery.

When disease occurs, the biochemical activity of cells begins to change. Cancer cells may multiply at a much faster rate and are more energetic than normal cells. As the disease progresses, this abnormal cellular activity begins to affect body tissue and structures, causing anatomical changes; Cancer cells may form a mass or tumour. Molecular imaging detects cellular changes early in the course of the disease. A variety of imaging agents are used to visualize cellular activity, such as the chemical processes involved in metabolism, oxygen use or blood flow. The imaging agent in the body accumulates in a target organ or attaches to specific cells. The distribution pattern of the agent helps to distinguish how well organs and tissues are functioning.

6.3 Radiosensitivity of children

Children are more radiosensitive as the organs and cells in children are undergoing constant self-renewal, therefore are more sensitive to radiation. Measurement of DNA synthesis by PET Radiopharmaceuticals that identify increased DNA synthesis can be used to identify increased cellular proliferation in tumours.

Children, due to increased mitotic activity and longer life expectancy, are more radiosensitive than middle-aged adult by a factor of up to 10 and girls are considered more radiosensitive than boys [12].

Radiosensitivity decreases with age, exhibiting lifetime attributable cancer mortality risks per unit dose as a function of age at a single acute exposure. This was estimated by the Committee on the Biological Effects of Ionizing Radiations (BEIR) [15, 16] and the International Commission on Radiological Protection (ICRP) [17].

Children are two to three times more susceptible to radiation for the development of leukaemia. Adults exposed to radiation during childhood have an increased likelihood of emerging breast or thyroid cancer.

The National Academy of Sciences BEIR V committee and the ICRP report 60 have estimated the lifetime cancer mortality risks per unit dose at a single acute exposure as a function of age. They have shown a rapid increase in lifetime risk with decreasing age at exposure (**Figure 7**).

This indicates that radiosensitivity decreases with age. Neonates are more radiosensitive than infants, infants are more radiosensitive than children and children are more radiosensitive than adolescents.

6.4 Radiation life-time risk

Radiation-induced cancers tend to appear at the same age as spontaneous cancers of the same type. So, it takes half a century or more to judge the impact of radiation exposure, especially when children are included in the exposed individuals.

Exposed to radiation individuals in their first decade of life, the risk is approximately 15% per Sv, while for adults in their late middle age, the risk drops to 1 or 2%/Sv.

DOI: http://dx.doi.org/10.5772/intechopen.105346

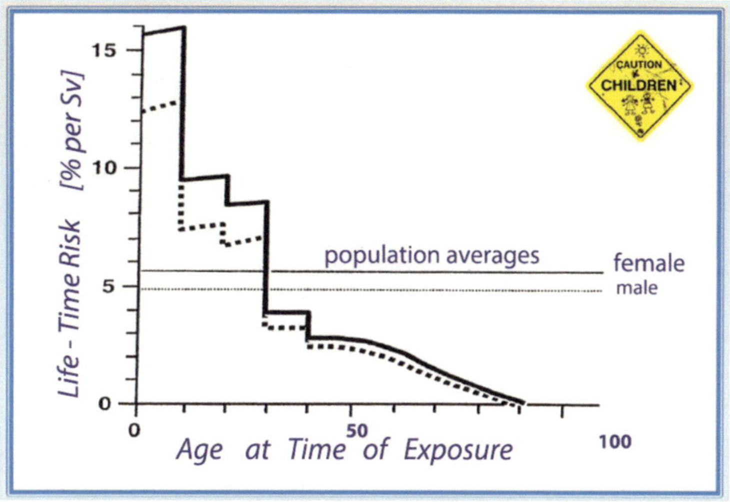

Figure 7.
The above risk estimate is an average for a population comprised of all ages. It is apparent that the risk varies dramatically with age (from Eric J. Hall, 2002), [15].

There is also a clear gender difference, especially at early ages, with girls being more radiosensitive than boys (**Figure 7**).

7. Paediatric patient-specific dosimetry

7.1 Individualized dosimetry

Paediatric Dose Reference Levels (PDRLs) must be established, especially in a national level and then effective dose estimations from images data can be obtained. The calculated PDRLs may help in the standardization of the appropriate activity in paediatric nuclear medicine.

Individualized dosimetry and iterative algorithms may reduce further the administered dose resulting in safer children's examinations.

To limit radiation exposure to children from diagnostic nuclear medicine procedures to the lowest levels consistent with quality imaging, a study has been established [18] to correlate administered activity/weight- to an effective dose in paediatric nuclear medicine imaging.

In radiopharmaceutical schedules for children, fractions of adult administered amounts and formulae based on the child's body parameters are used. Recommended activities could also be obtained by EANM dosage-card or North American Guidelines.

The paediatric administered activities are determined by the formula that reduces adult administered activity as:

Paediatric dosage [MBq] = (Child Weight [Kg] x Adult Reference Activity [MBq])/70.

Radiopharmaceutical dosages for five diagnostic radiopharmaceuticals (Tc99m-DMSA, Tc99m-DTPA, Tc99m-MAG3, Tc99m-MDP & I123-MIBG) were calculated for 100 paediatric imaging procedures and administered in terms of activity/kg.

Knowledge of physical and biological parameters is required for the calculation of the absorbed dose.

Absorbed dose is the average deposition of energy in the tissue from the administered radiopharmaceutical.

The radioactive elements, used in the diagnosis, are distributed to the human body following the rules of pharmacokinetics & pathophysiology (**Figure 8**).

The RADAR dosimetry program was used by Plousi et al [18] in order to estimate the effective dose per child per weight/age for various radioisotopes, with reduced reference adult activity being incorporated, **Figure 9**.

Weighting Factor of administered activities per weight (kg) were varied from (0.1–0.86%) for 3Kg weight of a neonate to 40Kg weight of an adult.

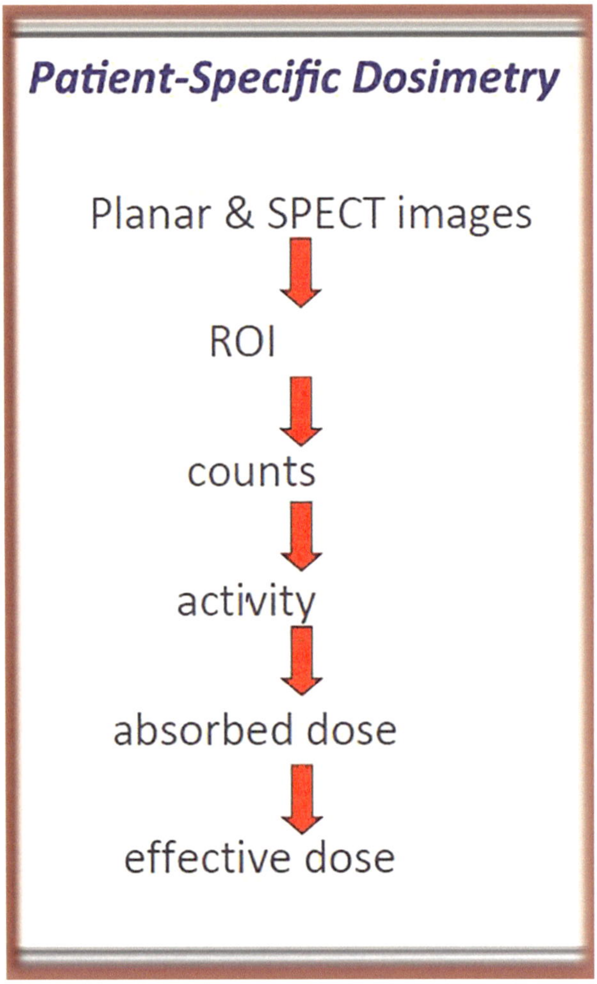

Figure 8.
The absorbed dose depends on: The administered activity. The active time of its stay in an organ. The parameters fixed in time, that is: radioisotope characteristics, shape and size of the radiating organ (source), the irradiated organ (target) and the distance and mass of the target) [18].

Effective Dose (mSv)

Radiopharmaceutical	newborn	1 year old	5 years old	10 years old	15 years old
Tc99m-DMSA	1.29	0.55	0.53	0.68	0.85
Tc99m-DTPA	0.6	0.38	0.54	0.66	1.36
Tc99m-MAG3	0.4	0.18	0.21	0.27	0.66
Tc99m-MDP	2.56	1.84	1.79	2.06	2.22
I123-MIBG	11.76	5.23	3.93	4.7	6.6

Figure 9.
The absorbed dose in neonates is extremely higher (because the activity is distributed over smaller volumes). Significant differences -about a factor of two and sometimes three in activity and effective dose were measured between underweight, average weight and corpulent children of the same age [18].

For neonates and infants' cases, a minimum administered activity is applied considering that the use of a fraction of the administered activity of adults would result in an uncompleted study. Planar whole-body and SPECT imaging studies were performed on a γ-camera equipped with a high-resolution collimator.

Regarding I123-MIBG, the lower limit [30 MBq for neonatal] and upper limit [110 MBq] was established to give the least effective dose with the best quality imaging.

For newborn cases, it is necessary to apply a minimum activity, as the activity calculated according to weight is less than the recommended minimum activity. When the suggested weight-based administered activities are used, the resulting effective doses range in ages 1–10 years old are, **Figure 10**.

Activities for Tc99m-DMSA for planar and 3D imaging are lowering as filtering and iterative reconstruction methods were used. In dynamic studies of paediatric patients, the SNMMI/EANM Guidelines for Diuresis Renography in infants and children were followed [19]. The lowest burden is estimated for Tc99m-MAG3.

Optimal protocols, with improved image reconstruction methods and advanced instrumentation, facilitate the dosage reduction and provide the maximum image quality at a minimum of effective dose [20].

A graphic relation of Administered Activity versus weight of all patient groups, from neonates to adolescents, is presented in **Figure 11A**.

In **Figure 11B**, a positive correlation of the effective dose (mSv) with patient ages (0–26years) is shown. No differences were observed between boys and girls of the same age [18].

7.2 Dosimetry aspects in hybrid molecular imaging applications in paediatric patients

Dose reduction in PET/CT and SPECT/CT studies with children can be achieved by optimized CT parameters and the administered activity of the radiopharmaceutical,

(0.59-0.85) mSv for Tc99m-MAG3
(0.74-2.22) mSv for I123-MIBG
(4.4 - 3.7) mSv for Tc99m-MDP

Figure 10.
Effective doses for ages between 1 and 10 years old.

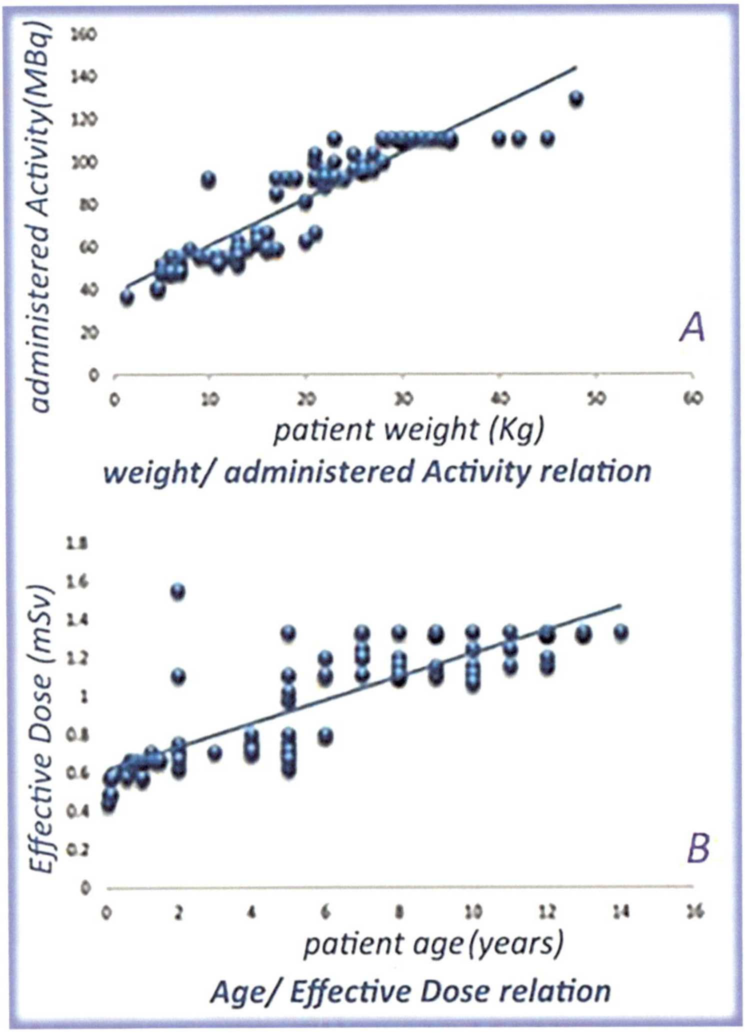

Figure 11.
(A) Administered Activity (MBq) to patient. Weight in Kg/from neonates to adolescents. (B) Positive relation of the effective dose (mSv) with patients' age (0–26 years). No differences were observed between boys and girls of the same age [18].

without compromising the diagnostic information needed for high-quality examination. Effective doses to the paediatric patient examined by PET/CT or SPECT/CT depend on the CT protocol of the accompanying CT scan. The co-registered CT scan can be optimized to meet the patient's diagnostic needs and may be performed either as a diagnostic-type CT scan or as an attenuation-correction only [21].

The hybrid molecular imaging examination by PET or SPECT and the CT should be acquired without child-patient movement. Attention to the respiratory phase during the CT imaging for PET/CT is also of a semantic point.

High-quality biokinetic data must be known for the calculation of dose estimates of new PET radiopharmaceuticals. Then, standardized dosimetry codes as OLINDA/EXM can provide information of doses to organs and effective doses [22].

In addition to the molecular imaging agents ^{18}F-FDG (PET) and ^{123}I-MIBG (SPECT) that are frequently used in children, other PET and SPECT imaging agents may have promise for molecular imaging in children.

- C-11-methionine by PET has been shown in several studies to better depict paediatric brain tumours when compared to FDG.
- SPECT/CT may be used to localize sites of abnormal I-131 uptake in thyroid cancer patients who are post-thyroidectomy or
- Tc-99m-HMPAO-labeled white blood cells may be used with SPECT/CT to localize areas of inflammation [21].

PET/MRI use in children with systemic malignancies may benefit from the reduced radiation exposure offered by PET/MRI. The effective dose of a PET/MRI scan is only about 20% that of the equivalent PET/CT examination. Simultaneous acquisition of PET and MRI data combines the advantages of the two previously separate modalities. One disadvantage of PET/MRI is that in order to have an effect, a significantly longer examination time is needed than with PET/CT. PET/MRI has turned out to be a stable hybrid imaging modality, which generates paediatric safe diagnostic studies [23].

7.3 Foetal doses from nuclear medicine examinations

Doses are provided for "early pregnancy" (dose to the nongravid uterus in the RADAR reference adult female model) and doses to the foetus at 3, 6, and 9 months of gestation (OLINDA/EXM 2.0 software).

Uncertainties in using these estimates for a specific subject are noteworthy, both in the physiology of the radiopharmaceutical kinetics and in the assumed geometry of the maternal and foetal organs [23].

Foetal whole-body doses from common nuclear medicine examinations in early pregnancy as well as at terms have been calculated by Russell and Stabin using ICRP 53 and ICRP 80.

For Example:

1. A pregnant woman at 4 months' gestation is administered 370 MBq of ^{18}FDG. The estimated foetal dose at 3 months is 4.8 mGy and at 6 months is 3.1mGy. An estimate of 5 mGy is reasonable and conservative [24].

Radiopharmaceutical	Procedure	Administered Activity	Early pregnancy	9 months
^{123}I-	Thyroid uptake	30 MBq	0.4-0.6mGy	0.3mGy
^{131}I-	Thyroid uptake	0.55	0.03-0.04mGy	0.15

Figure 12.
Foetal thyroid doses for 30MBq ^{123}I or 0.55 MBq ^{131}I in early pregnancy and at 9 months.

2. Foetal thyroid doses are much higher than foetal whole-body doses, 5–15 mGy/MBq for ^{123}I and 0.5–1.1 Gy/MBq for ^{131}I (**Figure 12**).

8. Discussion

8.1 Dosimetry in paediatric nuclear medicine: from acquisition to image processing, image gently

The radiation burden in nuclear medicine depends principally on the administered radiopharmaceutical properties and the biological parameters-pharmacokinetics properties of the radiopharmaceutical within the patient. So,

- Activity determination is one of the fundamental bases in nuclear medicine dosimetry.

- The calculation of effective dose in a paediatric patient varies also due to anatomical differences.

- The basic schema for dosimetry calculations involved in the MIRD formalism for radiopharmaceutical dosimetry.

- The ICRP models underlying the application of dosimetry.

- Imaging value in dosimetry is the best conversion of image data to absolute values of uptake.

To limit radiation exposure to children from diagnostic nuclear medicine procedures to the lowest levels with reliable qualitative imaging, a correlation of administered activity with weight-effective dose in radiopharmaceutical imaging is valued.

Administered activities in paediatric subjects are distributed over smaller volumes generating higher absorbed doses. In diagnostic examinations, fractions of adult administered amounts and formulae based on child's body parameters are used. Recommended activities could be obtained by EANM dosage-card or North American Guidelines—Paediatric dosage card.

Cumulated activity calculation from the time-activity curve will lead to a total number of disintegrations. Absorbed dose algorithms and image processing determine the radiation transport, energy deposition and the radiation burden of the subject; Advanced approaches such as Monte-Carlo modelling in nuclear medicine for imaging and dosimetry are successfully used.

For newborn subjects, it is necessary to apply the minimum dose, because the activity calculated according to newborn weight is less than the recommended minimum activity, resulting to worsen diagnostic imaging quality.

8.2 Optimization—Conclusion

Optimization in medical imaging is the balancing of the amount of ionizing radiation and image quality. The minimum radiation dose for the paediatric patient must assure that the image quality provides satisfactory information to meet the clinical requirement. Optimization involves both the imaging systems as testing and quality control as well as imaging body parameters and administered activity [25].

Optimal protocols, with improved image reconstruction methods and advanced instrumentation, facilitate the dosage reduction and provide the maximum image quality at a minimum effective dose. Optimization of imaging protocols and establishment of diagnostic reference levels achieve the goals of good quality images at reduced radiation doses. Standardized methods for performing dose calculations for radiopharmaceuticals by various steps in the process and models for calculating time–activity integrals as urinary bladder or intestines can be used [26].

In hybrid imaging PET/CT or SPECT/CT, deep learning-based reconstruction (DLR) may facilitate CT radiation dose reduction in children. Lower-dose DLR images were compared with standard-dose iterative reconstruction images. DLR use at 80-kVp results in substantial dose reduction with preserved or even improved image quality. So, the use of DLR allows greater dose reduction for paediatric CT than current image reconstruction techniques [27].

Clinical dosimetry in targeted radionuclide therapy in children supports the treatment decisions and should be a strong indication that treatment results are dependent on the absorbed dose delivered to the treated organ as well as to the critical organs.

Acknowledgments

D.A. Verganelakis gratefully acknowledges all support provided by the 'ELPIDA' Association of Friends of Children with Cancer and the Oncology Clinic "Marianna V. Vardinoyiannis" at Children's hospital "Aghia Sophia" in Athens.

Author details

Dimitris A. Verganelakis[1*] and Maria Lyra-Georgosopoulou[2]

1 Nuclear Medicine Unit, Oncology Clinic "Marianna V. Vardinoyiannis"—ELPIDA, Children's Hospital 'Aghia Sophia', Athens, Greece

2 Department of Radiology, Radiation Physics Unit, University of Athens, Athens, Greece

*Address all correspondence to: dimitris.verganelakis@gmail.com

References

[1] Treves TS, Lassmann M. International Guidelines for Pediatric Radiopharmaceutical Administered Activities, for the EANM/SNMMI Pediatric Dosage Harmonization Working Group, 2014

[2] Grant DF, Gelfand JM, Drubach LA, Treves TS, Fahey FH. Paediatric radiation doses for paediatric nuclear medicine studies: Comparing the North American consensus guidelines and the paediatric dosage card of the European Association of Nuclear Medicine. Radiology. 2015;**45**:706-713. DOI: 10.1007/s00247-014-3211-x

[3] Treves TS et al. Update: Image gently and nuclear medicine at 10 years newsline. N7-N9 The Journal of Nuclear Medicine. 2019;**60**(4):7N-9N

[4] Fahey FH, Bom HS, Chiti A, et al. Standardization of administered activities in paediatric nuclear medicine: A report of the First Nuclear Medicine Global Initiative Project, Part 1: Statement of the issue and a review of available resources. The Journal of Nuclear Medicine. 2015;**56**:646-651

[5] Fahey HF, Bom SH, Chiti A, et al. Standardization of administered activities in paediatric nuclear medicine, Part 2: A report of the First Nuclear Medicine Global Initiative Project. The Journal of Nuclear Medicine. 2016;**57**:1148-1157

[6] Khaman K, O'Reilly SE, et al. Re-evaluation of paediatric 18F-FDG dosimetry: Cristy-Eckerman versus UF/NCI hybrid computational phantoms. Physics in Medicine and Biology. 2019;**63**(16):165012. DOI: 10.1088/1361-6560/aad47a

[7] Stabin M, Farmer A. Meeting report molecular targeting probes—Radioactive and nonradioactive OLINDA/EXM 2.0: The new generation dosimetry modelling code. The Journal of Nuclear Medicine. 2012;**53**(suppl.1):585

[8] Bolch WE, Eckerman KF, Sgouros G, et al. MIRD Pamphlet No. 21: A Generalized Schema for Radiopharmaceutical Dosimetry-Standardization of Nomenclature. The Journal of Nuclear Medicine. 2009;**50**(3):477-484. DOI: 10.2967/jnumed.108.056036

[9] Stabin MG, Siegel JA. RADAR dose estimate report: A compendium of radiopharmaceutical dose estimates based on OLINDA/EXM Version 2.0. The Journal of Nuclear Medicine. 2018;**59**:154-160. DOI: 10.2967/jnumed.117.196261

[10] Vañó E, Miller DL, Martin CJ, Rehani MM, et al. Authors on behalf of ICRP, ICRP Publication 135: Diagnostic reference levels in medical imaging. Annals of the ICRP. 2017;**46**(1):1-144. DOI: 10.1177/0146645317717209

[11] Vassileva J, Rehani M, et al. Study to establish international diagnostic reference levels for paediatric computed tomography. Radiation Protection Dosimetry. 2015:1-11. DOI: 10.1093/rpd/ncv116;**165**(1-4):70-80

[12] Raissaki MT. Pediatric radiation protection. European Radiology Supplements. 2004;**14**:74-83. DOI: 10.1007/s10406-004-0011-7

[13] The JSNM. Japanese consensus guidelines for paediatric nuclear medicine. Annals of Nuclear Medicine. 2014;**28**:498-503

[14] Aerts A, Eberlein U, Holm S, et al. EANM position paper on the role of radiobiology in nuclear medicine. European Journal of Nuclear Medicine and Molecular Imaging. 2021;**48**:3365-3377

[15] Hall JE. Lessons we have learned from our children: Cancer risks from diagnostic radiology. Pediatric Radiology. 2002;**32**:700-706. DOI: 10.1007/s00247-002-0774-8

[16] BEIR VII. Health risks from exposure to low levels of ionizing radiation: BEIR VII—Phase 2 Committee to assess health risks from exposure to low levels of ionizing radiation. National Research Council. Washington DC: The National Academies Press, 2006. p. 424

[17] ICRP Publication 60, 1990 Recommendations of the International Commission on Radiological Protection Annals of the ICRP, April 1991| 202 pages| SAGE Publications Ltd

[18] Plousi A, Baka I, Gazeli E, et al. Dose reduction in paediatric nuclear medicine-paediatric diagnostic reference levels (PDRLs) in nuclear medicine imaging, EJNMMI 2012. Conference: 25th Annual Congress of the EANM;**2012**:OP092

[19] Majd M, Bar-Sever Z, Santos AI, et al. The SNMMI and EANM procedural guidelines for diuresis renography in infants and children. JNM. 2018;**59**(10):1636-1640. DOI: 10.2967/jnumed.118.2159211636

[20] ICRP. 1991 Recommendations of the international commission on radiological protection. Annals of the ICRP. 1991;**21**(1-3):1-201

[21] Gelfand MJ. Dosimetry of FDG PET CT and other molecular imaging applications in paediatric patients. Pediatric Radiology. 2009;**39** (Suppl. 1):S46-S56. DOI: 10.1007/s00247-008-1023-6

[22] Stabin MG. Radiation dosimetry of PET imaging. In: Basic Science of PET Imaging. Khalil MM, Springer, 2017. DOI: 10.1007/978-3-319-40070-9_3

[23] Hirsch FW, Sattler B, Sorge I, et al. PET/MR in children. Initial clinical experience in paediatric oncology using an integrated PET/MR scanner. Pediatric Radiology. 2013;**43**(7):860-875. DOI: 10.1007/s00247-012-2570-4

[24] Stabin MG. New-generation foetal dose estimates for radiopharmaceuticals. The Journal of Nuclear Medicine. 2018;**59**(6):1005-1006

[25] Stabin MG, Wendt RE, Flux GD. RADAR Guide: Standard methods for calculating radiation, doses for radiopharmaceuticals, Part 1-Collection of Data for radiopharmaceutical dosimetry. The Journal of Nuclear Medicine. 2022;**63**:316-322. DOI: 10.2967/jnumed.120.259200

[26] Stabin MG, Wendt RE, Flux GD. RADAR Guide: Standard methods for calculating radiation doses for radiopharmaceuticals, Part2-data analysis and dosimetry. The Journal of Nuclear Medicine. 2022;**63**:485-492. DOI: 10.2967/jnumed.121.262034

[27] Nagayama Y, Goto M, Sakabe D, Emoto T, et al. Radiation Dose Reduction for 80-kVp Pediatric CT using deep learning-based reconstruction: A clinical and phantom study. American Journal of Roentgenology. 2022;1-11. DOI: 10.2214/AJR.21.27255